Building Your Own
JavaScript Framework

Architect extensible and reusable framework systems

Vlad Filippov

BIRMINGHAM—MUMBAI

Building Your Own JavaScript Framework

Group Product Manager: Rohit Rajkumar

Publishing Product Manager: Himani Dewan

Book Project Manager: Aishwarya Mohan

Senior Editor: Rashi Dubey

Technical Editor: Simran Udasi

Copy Editor: Safis Editing

Proofreader: Safis Editing

Indexer: Manju Arasan

Production Designer: Alishon Mendonca

DevRel Marketing Coordinators: Nivedita Pandey and Namita Velgekar

First published: November 2023

Production reference: 1051023

Published by Packt Publishing Ltd.

Grosvenor House

11 St Paul's Square

Birmingham

B3 1RB, UK.

ISBN 978-1-80461-740-3

www.packtpub.com

To Erica, my guiding light, your inspiration propels me to embrace the unknown and expand my horizons. To Marina and Roman, for the belief and solid foundation of technological passions. And to all my dear friends, whose support over the years has been an anchor to my journey.

– Vlad Filippov

Foreword

It's hard to overstate JavaScript's impact on the world. What was once seen as a toy scripting language, designed in just 10 days, today enjoys the position as the most popular programming language on GitHub and has evolved into a modern, mature language. The abundance of JS frameworks that have emerged in recent years is a testament to the language's versatility and longevity. From Angular to React, Vue.js to Svelte, these frameworks have revolutionized the way we approach frontend web development. Beyond the frontend, JavaScript powers some of the most popular native desktop and mobile apps enabled by projects such as Electron and React Native and is even being used to power the SpaceX Dragon 2 console *in space*. Complementing software design advances, there are dozens of JavaScript testing frameworks and developer productivity solutions, each with its own advantages that scale from a simple command-line interface to large-scale enterprise offerings. The varied JavaScript frameworks have enabled millions of developers to be efficient and effective in their work.

I met Vlad in person for the first time at a jQuery conference in Cambridge, Massachusetts, more than 14 years ago. Even before that, Vlad and I first collaborated on providing support to the jQuery frontend library. What began for me as a way to learn more about jQuery and JavaScript turned into a substantial collaboration with Vlad working on the Firefox web browser, being part of the web standards and JavaScript community, speaking at conferences across the world, and hacking on open source projects. Over the years, I've remained impressed at Vlad's energy, creativity, knowledge, and passion for open source as he developed multiple internal frameworks at Mozilla, maintained the Grunt.js task runner, and contributed to many, many projects related to JavaScript infrastructure and developer productivity. This book is a testament to Vlad's expertise in open source JavaScript development and abilities to teach complex topics. You will find a blend of JavaScript framework development knowledge in this book, covering a wide range of topics, from the basics of how a framework is beneficial and how to build one to advanced concepts such as extensible components and framework architecture.

Whether you are a curious newcomer to the world of web development or a seasoned professional seeking to expand your horizons, you will find relevant knowledge in this book. As you follow along building your own framework and applying the lessons in your daily life, you'll gain the skills needed to meet your own varied goals, whether they be joining an open source framework community, learning to become a more well-rounded engineer, or refactoring an internal legacy application using the patterns found in this book, or just building something for fun late at night.

Mike Taylor (@miketaylr)

Engineering manager, Privacy Sandbox Team at Google

Contributors

About the author

Vlad Filippov is an experienced full stack developer who has been deeply immersed in software for more than a decade. His work spans a wide range of domains, from API and web development to mobile applications and 3D visualizations. Vlad is currently working at Stripe, creating financial infrastructure for the internet. Previously, Vlad spent over six years working on the Firefox web browsers and cloud services at Mozilla.

Vlad holds a computer science specialist degree from the University of Toronto and has presented at over a dozen conferences around the world. His passion lies in web applications and open source projects, primarily focusing on JavaScript apps and development tooling. In the brief moments he is not coding, Vlad enjoys playing tennis and running.

About the reviewer

Deepak Vohra is the technical reviewer of numerous (25+) books for various publishers including APress Media, Manning Publications, and Packt Publishing. Deepak is the author of 20+ books on various topics including Java, Docker, Kubernetes, and Apache Hadoop.

Table of Contents

6

Building a Framework by Example   105

7

Creating a Full Stack Framework   123

Part 3: Maintaining Your Project

Preface

This book is about crafting, understanding, and maintaining JavaScript frameworks – the critical building blocks of many modern web applications. The ecosystem of software frameworks, particularly JavaScript ones, has been highly interesting for many developers in the past 10-plus years. In addition, JavaScript has been arguably the most influential and ubiquitous programming language, enabling millions of critical web applications and services. Powered by the vibrant community of developers and innovative companies, the development of JavaScript frameworks has since emerged as a dynamic and exciting space, with opportunities for innovation and exploration. Each new framework brings its own philosophy and approach, representing a unique set of solutions to common web application and web services challenges.

It is important to note that this book is not solely for those planning to develop their JavaScript frameworks from scratch. While it provides valuable wisdom in creating a well-maintained system, its primary goal extends beyond that. It is also designed to refine your understanding of frameworks operating under the hood, refining your JavaScript or software developer skills. It encourages you to think from a framework's perspective, thus fostering a more profound comprehension of reusable software architectures. Whether you're an experienced developer looking to delve deeper into your toolkit or a newcomer seeking to understand the framework landscape, this book will be advantageous to your career.

Who this book is for

If you are a JavaScript novice or an expert who wants to dive deep into the world of JavaScript frameworks, we've got you covered. This book introduces you to the history of frontend frameworks and guides you through the creation of your own framework. To learn from this book, you should have a good understanding of the JavaScript programming language and some experience with existing software frameworks.

What this book covers

Chapter 1, *The Benefits of Different JavaScript Frameworks*, provides a detailed introduction to the world of JavaScript frameworks and their growth over the last decade.

Chapter 2, *Framework Organization*, provides insights into how frameworks are structured and their organizational patterns.

Chapter 3, *Internal Framework Architecture*, examines the core technical architectural patterns of various existing frameworks.

Chapter 4, *Ensuring Framework Usability and Quality*, explains a series of techniques and tools that enable developers to build quality frameworks focused on usability and ease of use.

Chapter 5, *Framework Considerations*, explores considerations that revolve around project goals, the technical problem space, and architectural design decisions.

Chapter 6, *Building a Framework by Example*, explains how to develop a simple JavaScript testing framework based on the patterns and best techniques.

Chapter 7, *Creating a Full Stack Framework*, offers practice in developing a new application framework that will enable developers to build large and small web applications.

Chapter 8, *Architecting Frontend Frameworks*, specifically focuses on the frontend components of a full stack framework.

Chapter 9, *Framework Maintenance*, discusses topics relating to the framework release process, the continuous development cycles, and the long-term maintenance of large framework projects.

Chapter 10, *Best Practices*, explains several vital topics around general JavaScript framework development and takes a glimpse into the future of frameworks in this ecosystem.

To get the most out of this book

You will need to have a good understanding of the JavaScript programming language and have some experience with existing software systems such as React.

Software/hardware covered in the book	Operating system requirements
ECMAScript 2023	Windows, macOS, or Linux
TypeScript	
Node.js 20	

Follow the guidelines to install Node.js version 20+ on your machine to be able to interact with the code files. The installers for the Node.js runtime can be found at `https://nodejs.org/en/download`.

If you are using the digital version of this book, we advise you to type the code yourself or access the code from the book's GitHub repository (a link is available in the next section). Doing so will help you avoid any potential errors related to the copying and pasting of code.

Download the example code files

You can download the example code files for this book from GitHub at `https://github.com/PacktPublishing/Building-Your-Own-JavaScript-Framework`. If there's an update to the code, it will be updated in the GitHub repository.

We also have other code bundles from our rich catalog of books and videos available at `https://github.com/PacktPublishing/`. Check them out!

Conventions used

There are a number of text conventions used throughout this book.

`Code in text`: Indicates code words in text, database table names, folder names, filenames, file extensions, pathnames, dummy URLs, user input, and Twitter handles. Here is an example: "For example, if you are planning to add API routes to build an API in the Next.js project, they must be mapped to an `/api/` endpoint."

A block of code is set as follows:

```
pages/index.vue
<template>
  <NuxtLink to="/">Index page</NuxtLink>
  <NuxtLink href="https://www.packtpub.com/" target="_blank">Packt</
NuxtLink>
</template>
```

When we wish to draw your attention to a particular part of a code block, the relevant lines or items are set in bold:

```
<root framework directory>
  | <main framework packages>
    + <core framework interfaces...>
    + <compiler / bundler>
  | <tests>
    + <unit tests>
    + <integration and end-to-end tests>
    + <benchmarks>
  | <static / dynamic typings>
  | <documentation>
  | <examples / samples>
  | <framework scripts>
  | LICENSE
  | README documentation
  | package.json (package configuration)
  | <.continuous integration>
  | <.source control add-ons>
  | <.editor and formatting configurations>
```

Bold: Indicates a new term, an important word, or words that you see on screen. For instance, words in menus or dialog boxes appear in **bold**. Here is an example: "After pressing the **Debug** button, you will be presented with a context menu of all the available scripts in the project."

> **Tips or important notes**
> Appear like this.

Get in touch

Feedback from our readers is always welcome.

General feedback: If you have questions about any aspect of this book, email us at customercare@ packtpub.com and mention the book title in the subject of your message.

Errata: Although we have taken every care to ensure the accuracy of our content, mistakes do happen. If you have found a mistake in this book, we would be grateful if you would report this to us. Please visit www.packtpub.com/support/errata and fill in the form.

Piracy: If you come across any illegal copies of our works in any form on the internet, we would be grateful if you would provide us with the location address or website name. Please contact us at copyright@packt.com with a link to the material.

If you are interested in becoming an author: If there is a topic that you have expertise in and you are interested in either writing or contributing to a book, please visit authors.packtpub.com.

Share Your Thoughts

Once you've read, we'd love to hear your thoughts! Scan the QR code below to go straight to the Amazon review page for this book and share your feedback.

https://packt.link/r/1804617407

Your review is important to us and the tech community and will help us make sure we're delivering excellent quality content.

Download a free PDF copy of this book

Thanks for purchasing this book!

Do you like to read on the go but are unable to carry your print books everywhere?

Is your eBook purchase not compatible with the device of your choice?

Don't worry, now with every Packt book you get a DRM-free PDF version of that book at no cost.

Read anywhere, any place, on any device. Search, copy, and paste code from your favorite technical books directly into your application.

The perks don't stop there, you can get exclusive access to discounts, newsletters, and great free content in your inbox daily

Follow these simple steps to get the benefits:

1. Scan the QR code or visit the link below

https://packt.link/free-ebook/9781804617403

2. Submit your proof of purchase
3. That's it! We'll send your free PDF and other benefits to your email directly

Part 1:
The Landscape of
JavaScript Frameworks

The first four chapters delve into the ecosystem of existing frameworks. This approach gives a unique and diverse perspective into the current technological landscape of JavaScript frameworks. These chapters emphasize the facets of structure, design, and methodologies for guaranteeing excellence and developer-friendliness in such projects. The knowledge accumulated here empowers developers to understand the broader context and underlying principles of framework development, with the goal of creating a solid foundation.

In this part, we cover the following chapters:

- *Chapter 1, The Benefits of Different JavaScript Frameworks*
- *Chapter 2, Framework Organization*
- *Chapter 3, Internal Framework Architecture*
- *Chapter 4, Ensuring Framework Usability and Quality*

1
The Benefits of Different JavaScript Frameworks

It has been over 25 years since JavaScript was first introduced into our web browsers. Since then, this technology has vastly changed how we interact with websites and applications, how we build APIs for backend systems, and even how we communicate with hardware platforms. JavaScript has become one of the most popular programming languages on the planet. To this day, JavaScript's pace of evolution and rapid change is a popular topic of conversation among developers – it is a source of excitement and innovation. As a programming language, JavaScript has been ranked as the most popular among developers in the last 10 consecutive years and has been the key to client-side scripting for 98% of all websites. We cannot underestimate how much JavaScript and its closely related ECMAScript standardization have enabled the web to become the platform to host the next generation of software that can be accessed by billions of people. With these technologies, millions of businesses and individuals can easily build great applications, creative experiences, and complex software solutions. In many ways, the web platform has the potential to be the most vibrant and friendly developer ecosystem in the whole world.

JavaScript frameworks are the straightforward way millions of web developers build projects today. Due to their popularity and ease of use, frameworks allow developers to quickly make product ideas come to life without unnecessary overheads. Without the framework systems that we have at our disposal today, the web would not have been able to compete with other development platforms.

In this book, we will study the vast ecosystem and expand our knowledge to become confident in creating and maintaining our own self-developed frameworks. Developing the skill to build a framework or extend existing ones comes with the benefit of becoming an impactful domain expert in frontend and backend projects.

As part of becoming experts in JavaScript frameworks, we need to get a sense of the core components and tools of the web development workflow. In this first chapter of the book, we will take a look at how web development evolved, how frameworks changed the landscape of working with JavaScript, and what the ecosystem currently has to offer.

We will cover the following topics:

- The emergence of JavaScript frameworks
- The evolution of code bases
- Types of frameworks in JavaScript and their benefits
- My experiences with frameworks

Technical requirements

This book has an accompanying GitHub repository at `https://github.com/PacktPublishing/Building-Your-Own-JavaScript-Framework`. In each chapter, we will point to the relevant directories in this repository. Feel free to clone or download the repository as a ZIP file.

You need a desktop or a laptop computer with internet access and a terminal application to install and run the code from this repository. We shall also be utilizing Node.js to run some of the parts of the repository. Node.js is an open source, cross-platform, backend JavaScript runtime environment that runs JavaScript code outside a web browser. The installation for Node.js can be found at `nodejs.org`. For the code from the repository, you can use any environment that supports a terminal and runs Node.js, such as Windows, macOS, and most varieties of Linux.

The emergence of JavaScript frameworks

As JavaScript progressed and evolved, the innovators who were heavily involved in the language, both companies and individuals, started writing software libraries to help solve the everyday burdens of web application architectures. The initial focus of the most basic JavaScript libraries was to provide singular features, interactivity, and add-ons, which progressively enhanced the web page. At that time, JavaScript gave life to a static page with its interactive components – simple examples that always come to mind are tiny scripts, which enabled creative button effects and mouse cursor effects. In many cases, these scripts were separate from the site's core functionality and were not essential to allow users to interact with the content. Since the inception of small libraries, these have paved the way for the complex framework systems that we have today. The frontend technology quickly evolved, and now, developers are much more accustomed to megabytes of scripts powering the frontend code.

JavaScript libraries were the next step in the web development evolution, assisting with cross-browser quirks, complex visual effects, network requests, and web page layout. With the use of these libraries, developers were able to have the cross-browser development challenges under control. CSS started to catch up with its focus on layout features and cross-browser standards, improving the web's styling features. Developers finally began introducing structure and well-crafted systems into web development.

The time has come to finally focus on building scalable and opinionated software on the web, and this is where we began to see glimpses of complex software paradigms introduced into large websites and web applications. Companies and larger enterprises started treating the web as a serious application platform, which led to a few prominent projects written in JavaScript and compiled to JavaScript from languages such as Java. Tracing back to late 2009, we see the first iterations of **Model-View-Controller** (**MVC**) frameworks built entirely with HTML, CSS, and JavaScript. This MVC model allows more extensive projects to stay organized, enriches the development workflow, and opens up the world of frontend development to developers who expect a much more structured approach when writing software. The MVC model fit web applications well enough to spawn a renaissance in framework development.

Many development hours were invested into connecting the mechanisms between the JavaScript engines and the browser web APIs. In *Figure 1.1*, we see a simplified view of how this interaction happens:

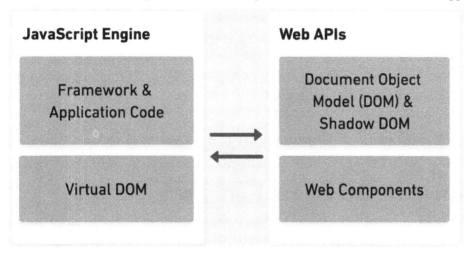

Figure 1.1: Interaction between the JavaScript engine and web APIs

The framework code and its internal technologies, such as the virtual DOM, use the DOM and its components to make the web application experience possible. The web has its own approach to the MVC architecture, with DOM and web API events interacting with Controllers defined in JavaScript. The Controllers interface with Views written in HTML or templated HTML. Furthermore, in this paradigm, the components of the applications utilize a Model to model the data within them. Using this approach, we can later communicate with backend services to retrieve the data in a particular manner.

Each new JavaScript MVC framework tried to perfect its implementation or approach in various ways. About five years after the first MVC frameworks appeared and gained popularity, several new paradigms that focused on the **observer** software design pattern started gaining traction in the JavaScript community. This observer approach is a software design pattern in which an object maintains a list of its dependants, called observers. The object notifies the observers automatically of the state changes within it. At that time, **Flux** came about, which is an application architecture that concentrates on simplifying the hurdles faced within MVC. The burdens included dealing with views constantly needing to interact with the models, hard-to-debug, deeply nested logic, and the need for adequate testing solutions of complex applications.

In the observer pattern, we define subjects that include a collection of observers informing about state changes. The Flux architecture expanded this existing pattern to fit better with applications built on the web. In the case of the Flux pattern, it consists of **stores** interacting with the state of a component. These stores get notified by a **dispatcher** based on data coming from **actions** that a user in the **view** took. Many JavaScript frameworks started adapting this pattern, ultimately simplifying how engineers structured their applications while still enforcing a set of applicable rules to keep the separation of concerns intact. The software patterns in all of these frameworks provide a clear separation of concerns between the interface, data model, and the application logic that integrates them. The Flux-based JavaScript frameworks introduced new concepts that emerged from the known MVC patterns. Still, both the MVC and Flux approaches focused on the principle of the separation of concerns in application development.

Furthermore, while simplifying ideas brought to light by Flux, a library called Redux inspired the next generation of frameworks to switch their approach to application state management. Instead of the Flux dispatchers, Redux frameworks rely on a single store with pure **reducer** functions, taking the current state and returning an updated state. Even today, frontend patterns are still maturing, and building for the web platform is becoming increasingly easier.

While there's a lot to mention with regard to frontend technologies, JavaScript has also made a great impact in places outside of web browsers. We cover those areas in the next section.

Frameworks outside the web browser

Another monumental event during the appearance of the first frontend frameworks was the emergence of a new open source runtime called Node.js. Node.js allowed developers to use JavaScript to produce server-side scripts, deploy backend systems, build developer tools, and, more importantly, write frameworks using the same language as the one from the web browser. The unique combination of having JavaScript on both sides of the software stack created immense possibilities for software developers. This runtime has since spread into many directions beyond software applications, with frameworks for desktop application development, hardware I/O solutions, and much more.

JavaScript-built frameworks enabled the web platform to become one of the most important technologies within reach of billions of people. It's almost impossible to imagine starting a new project without relying on the consistency and amiability of using a framework, and even the smallest tasks benefit significantly from using a cohesive and opinionated structure. However, even with the fast-paced evolution of the language and how we build web projects, it took quite a bit of time for JavaScript frameworks to emerge as fully encapsulated platforms that can aid developers in producing efficient applications.

JavaScript prevailed through the rise of mobile platforms, with multiple frameworks being created for mobile and existing systems integrating mobile benchmarks into their release process. The optimizations got to the hardware level, having the **ARM** (`arm.com`) processor architecture introduce optimizations to improve JavaScript performance in data type conversion, resulting in performance boosts for many JavaScript applications. That is quite a journey for a scripting language that started with small scripts on plain web pages.

Today, we can create fully fledged applications and services using the web platform by combining the power of web APIs, the JavaScript language, and technologies such as progressive web apps, using the frameworks that bring it all together. It is a fantastic time to start traversing the world of these JavaScript systems and using them to our advantage.

Now that we have an overview of how web development evolved, let's take a look at how the code bases have changed over time.

The evolution of code bases

While learning about frameworks, it is fascinating to reflect on how building for the web has changed over time. This exploration helps us understand why we build web applications the way we do today and helps us learn from historical shifts. It also allows us to be more mindful concerning framework usability and development decisions when we take on large projects. As technology pushes forward, the requirements and expectations around how websites and web applications are built drastically change. Depending on how long someone has been involved in web development, they either experienced many rapidly evolving shifts to how the code bases are structured or were lucky enough to avoid the times when the tooling and the process were highly tedious.

Initially, the code bases comprised isolated frontend components stitched together, consisting of code repetition and mixes of coding patterns. Code organization, usage of software development patterns, and performance optimizations were not a primary focus for developers. The web application deployment process used to be rudimentary as well. In many cases, the websites were manually updated and did not use source control or version tracking. Testing was also highly manual and would only exist in a few projects that were large enough to enable it. This was before deployment pipelines with continuous integration, deployment automation, and advanced testing infrastructure, rigorously verified every change. There used to be a time when developers had to optimize their CSS selectors for performance reasons.

Luckily, productivity and workflows rapidly started to improve once the industry started focusing more on building complex applications. Today we have source control, we have a myriad of testing and deployment tools to choose from, and we have established software paradigms that considerably improve our lives as developers and vastly improve the quality of the projects we build. Improvements to JavaScript engines unlocked new pathways for frameworks, and enhancements to web browsers fixed slow DOM interactivity with techniques such as the **virtual DOM**, **Shadow DOM**, and **Web Components**. These days, frontend frameworks have a better client platform to target as well, and the more established and improved web standards make it possible to perform much more complex operations. For example, with the help of WebAssembly (`webassembly.org`) standards, we can now run low-level code with improved performance, all within the browser.

As part of all these developments and growth in popularity, the web application development workflow got a lot more complex in many ways. Almost at every point of interaction with a web application project, there is a tooling set designed to improve our workflow. Some examples of this would be Git source control, various pre- and post-processors of our files, code editors with plugins, browser extensions, and many more. Here we have an example that illustrates the key components of a modern web application code base structure, in this case, generated by **SvelteKit**:

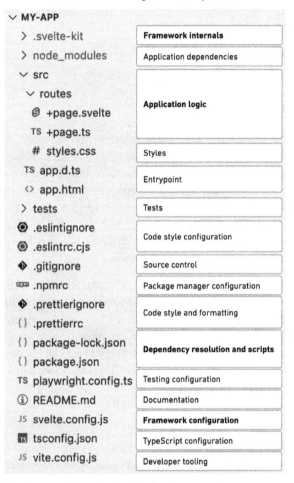

Figure 1.2: SvelteKit code base structure

We will go over SvelteKit later in the *Frameworks that use React* section of this chapter, and even if you have never used Svelte, this project file tree will look very familiar if you work with other frameworks. This dynamic structure of tools enables flexibility when it comes to switching out certain functionality. For example, *Prettier* can be substituted for another code formatting tool if need be, while the rest of the project structure remains the same and functions as it was.

With the establishment of the first frameworks in JavaScript, we experienced the introduction of a build step into our projects, which meant that either external or bundled tooling would help run or build the application. Today, this build step, popularized by **Webpack** or **esbuild**, is almost impossible to avoid. As part of this build step, we fetch application dependencies using package managers, process CSS, create code bundles, and run various optimization steps to make our app run fast and consume the least bandwidth. The ecosystem also introduced JavaScript transpilers, which are a type of source-to-source code compiler. They are used to take one particular syntax, which could consist of more modern features or include additional features, and convert them to be compatible with broadly accepted JavaScript syntax. Transpilers, such as **Babel**, began to see everyday use, integrated with the build step in many projects; this pattern generally motivated the use of the latest language features while also supporting old browser engines. These days, transpilation and build steps apply to files beyond JavaScript, as well as files such as CSS and specific templating formats.

Integrating with the build step are the package managers, such as npm or yarn, which play an essential role in resolving project dependencies. If you want to bootstrap a workflow with a framework, you will likely rely on the package manager to initialize the framework structure and its dependencies. For new projects, it is almost impossible to have a sensible framework workflow without using a package manager or some form of dependency resolution. As the project grows, the package manager facilitates the organization of newer dependencies while keeping track of updates to modules that are already in use. These days text editors, such as Visual Studio Code and IntelliJ WebStorm, adapt to our code bases and provide excellent tooling to enable source control of our code. The editors rely on built-in features and external plugins that encourage better formatting, easier debugging, and framework-specific improvements.

The code bases will keep changing as technology develops further, and the tools will keep improving to enable us to develop applications more quickly. Regarding the framework organization, we can expect higher levels of abstractions that simplify the way we do web development. Many programming languages, such as Java and Swift, have pre-defined development workflows encapsulating all aspects of development. JavaScript code bases so far have been an exception to these rules and allowed for high levels of flexibility. This trend is going to continue for many more years as the rapid pace of tooling and innovation in web development is not slowing down at all.

Now that we understand how the JavaScript ecosystem has evolved and how codebases have changed over time, let us explore what JavaScript frameworks offer in terms of frontend, backend, testing, and beyond.

Types of JavaScript frameworks and their benefits

Though it is challenging to compare all the subtle differences of every framework out there in the ecosystem, we can cover several frameworks that significantly impact the developer community or offer a unique approach to solving a particular problem. The expanded knowledge of the tools helps us notice specific patterns in these frameworks regarding the different strategies for developer experience and feature sets.

There are still ways to build apps and websites without frameworks, but many developers prefer to use established and opinionated frameworks even with the given overhead and learning curves. If you follow the JavaScript community, you will find that it is always passionately abuzz with discussions around frameworks, so let us dive deeper into the needs and benefits of framework use.

The frameworks provide good levels of abstraction to write high-level code without rewriting low-level functionality. Developers can be much more involved in business and product logic and iterate faster on new features. To give an example, up until recently, writing the code to make an asynchronous web request with proper error handling was a very time-consuming task without the aid of a good abstraction. Now that we have the Fetch API (`fetch.spec.whatwg.org`), this is a much easier endeavor, but Fetch is only part of the story, so the rest of the web APIs, especially the ones from earlier times, still benefit from good abstractions. In cases where we choose to write low-level code, it is a much better approach to find ways to write that code within the framework boundaries. This way it is tested and maintained within the primitives of the framework. This yields the benefits of avoiding extra maintenance and ensuring all the usages of that code are still behind sensible abstractions. Some backend frameworks approach this by providing extensible interfaces to hook into framework internals through plugins or extending the default behavior.

Developing software with groups of people is a challenging endeavor, so small and large teams can benefit from integrating a framework into the engineering workflow. The provided structure of abstractions generally results in much more well-architected systems, given the limits of how developers can write high-level components. The key benefit is enabling everyone involved in the task to understand the code base better and conveniently spend less time deliberating refactors and adding new code.

Now that we have our abstracted high-level code, we can cherish another benefit of frameworks – the performance optimizations they enable. Writing performant code that works in all provided use cases takes skill and takes away significant time from the project at hand. Even the most knowledgeable developers would only be able to come up with good enough solutions in a short amount of time. With frameworks, especially open source ones, you benefit from many minds put together to solve performance bottlenecks, overcome typical encumbrances, and continue to benefit from improvements as the framework develops. The performance benefits come from optimized low-level and well-structured high-level components; notably, some frameworks will guard against code that will slow down the application.

Frameworks make integrating with external systems, such as databases, external APIs, or specific components, easier. For instance, some web frameworks can integrate directly with the GraphQL data query language, simplifying backend systems' interaction. It's not just the ease of use, but also these integrations enable safe interaction with components such as databases, which helps avoid problematic queries that can be slow or harmful to execute. For frontend projects, it is important to always keep up with the latest web standards, and this is where frameworks provide another integration benefit.

Finally, as with all software, support plays an important role. Another reason a project may use an established framework is the available support channels through paid, volunteer, and open source help. The shared knowledge of these systems enables developers to help each other build systems and makes it easier to hire new developers who are familiar with these existing systems.

As we see, frameworks benefit us in countless ways – let us recap with these exact reasons. Here's what frameworks allow us to do:

- Focus on business logic and writing high-level code

- Write less code and follow code conventions defined by the framework

- Benefit from performance gains and rely on future optimizations

- Develop the project with good architecture, abstractions, and organization

- Easily integrate with external systems such as databases and external APIs

- Ability to rely on security fixes, audits, and patches

- Improve developer workflow using framework-specific toolings, such as text-editor integrations and command-line utilities

- Ability to debug issues easily by relying on detailed error messages and consistent logging

- Rely on external support from framework authors and the community

- Hire more developers who are already accustomed to working with the framework of our choice or with similar experience

- Develop better user experiences by leveraging the framework feature set

While a lot of JavaScript frameworks focus on the developer experience, the user experience can sometimes suffer from the overhead of these systems. This is usually relevant in frontend projects – an example of this would be loading a complex web application on budget mobile devices. In backend systems, this can be seen when the APIs are not able to keep up with request load and reliably scale with traffic spikes.

Even if the systems are skillfully built in both of these cases, the framework of choice might not be optimized to cover all use cases. I believe the next iteration of the framework ecosystem will largely focus on the user experience aspects, which means making load times faster, shipping less JavaScript over the network, and ensuring the web applications we create work seamlessly on all platforms. In the following sections, we will examine some of the most popular frameworks that enable these benefits for web application developers.

Frontend frameworks

Since JavaScript frameworks originated in the browser, let us look at modern frontend frameworks as our first exploration.

Ember.js

Suppose we trace the roots of the first JavaScript frameworks through the origins of libraries such as `Prototype.js`, `jQuery`, and `script.aculo.us`. In that case, we will eventually arrive at **SproutCore**, a framework used by Apple and a handful of other companies to build some of the most complex web experiences many years ago.

Today this early SproutCore project has influenced the **Ember.js** framework. Ember continues to be a highly opinionated piece of software that allows us to build applications with defined components, services, models, and a powerful router. Like many frameworks we will discuss in this chapter, Ember comes with its own command-line tooling, which helps developers quickly get started on the basics of the application and later generate more code quickly as the project scope grows. The usefulness of the provided framework tooling is immense. The CLI encapsulates the code generation steps and enables a way to run common framework commands, such as running tests or serving application files. With Ember, developers get a complete set of tools such as auto-reload, browser developer tooling, and a package called Ember Data, which helps manage the API-to-model relationship through adapters and serializers. Ultimately, Ember has a steeper learning curve than other frameworks, but its highly opinionated concepts guide developers toward highly functional web applications.

Angular

Angular is another framework with a large following. With TypeScript at its core, it is often used as a subset system for other full stack web frameworks. Angular provides its opinionated approach to component-based architecture. Angular has a complex history of rewrites but is now a more streamlined project with a stable feature set. Angular's template syntax extends HTML by adding expressions and new attributes. At its core, it uses the pattern of dependency injection. The latest versions of this framework offer a variety of binding techniques, including event, property, and two-way binding.

Vue.js

Vue.js, also written in TypeScript, was created by borrowing the good elements of Angular. Developers love Vue's simplicity within its component system, syntax, and general ease of use. It utilizes the **Model–View–Viewmodel** (**MVVM**) pattern, where a View communicates with a ViewModel using some data binding technique. In the case of Vue.js, for its data, it uses different techniques through HTML classes, HTML elements, and custom binding element attributes to achieve this. The purpose of the given ViewModel is to handle as much of the View's interaction logic and be the middle structure between the presentation logic and the application's business logic. Besides using HTML authoring, Vue has the **Single-File Component** (**SFC**) format (`vuejs.org/api/sfc-spec.html`) to encapsulate all aspects of the components – scripts, styling, and templating into one file. The SFC happens as part of the build step and helps the components avoid runtime compilation, scopes the CSS styles to the component, enables Hot Module Replacement, and much more.

About TypeScript

TypeScript is a superset of JavaScript, enabling features such as static typing and language extensions in many JavaScript environments. In recent years, TypeScript has become popular among framework authors and developers. It is also widely supported by many code editors and IDEs. Initially released in 2012 and inspired by ActionScript and Java, TypeScript is a superset of JavaScript, enabling features such as static typing and language extensions in many JavaScript environments. It helps catch errors at compile time instead of runtime error handling. Files with the file extensions `.ts` and `.tsx` are TypeScript files that must be compiled to JavaScript to be used in most environments.

Frameworks that use React

These days, we hear about **React** a lot; even though it is a user interface component library by itself, it has become the cornerstone for many frontend frameworks, such as **Gatsby**, **Remix**, **Next.js**, and others. As part of its introduction, React also debuted **JSX**, its own set of extensions to JavaScript, making it possible to define the components in a similar-looking syntax to HTML. For instance, the static site framework Gatsby relies on React's state management and the nested component architecture to compose its web pages. With Gatsby, developers can multiplex data, utilizing GraphQL, from content management systems, e-commerce sources, and other places.

Following along our React route, we get to Remix, which bundles a full stack solution with features for both the server and the client, plus a compiler and a request handler. Remix provides solutions for the View and Controller aspects of the application and relies on the Node.js module ecosystem for the rest, giving flexibility to the developers who need custom solutions from project to project. Based on the experience of creating and maintaining the `react-router` project for many years, the creators of Remix were able to come up with powerful abstractions while taking advantage of the browser's web APIs instead of investing in new concepts. To give an example, if you choose Remix for your project, you will find yourself using web standard APIs more than some of the other frameworks.

Next.js is our next React-based framework, which extends the use of the React component architecture as well by bringing it to the server with its built-in server rendering. The server-rendered components allow for a pre-rendered page to be sent to the client, resulting in the client only spending resources on initializing the interactive components. The framework provides the concept of pages, which allows for simpler routing implementations with lazy loading and enables automatic code-splitting. Combining all these features results in a fantastic user experience with fast loading times. In addition, the deployed applications rank highly when indexed by search engines, a feature that makes this framework stand out.

While talking about React frameworks, it is worth mentioning **Solid.js**. It's a newer library that creates frontend interfaces. Solid's benchmarks outperform React and others. It uses features such as JSX, but with a few key differences. With Solid, there is no virtual DOM and no concept of hooks. Instead, it relies on the pattern of **signals** to update the real DOM nodes, while utilizing reactive primitives. As part of Solid's approach, it offers the **SolidStart** app framework, which is very comparable to Next.js. It consists of core support components – *router*, *session*, *document*, *actions*, *data*, *entrypoints*, and *server* – these are integrated together as part of SolidStart.

SvelteKit

Like SolidStart, there is also **SvelteKit**, a new framework powered by the **Svelte** user interface library. SvelteKit framework generator script assembles a project skeleton, which helps you quickly get started and gives you full flexibility over the configuration of the compiler and the TypeScript settings. When setting up SvelteKit, we can use JavaScript with JSDoc formatting or TypeScript directly to write frontend applications. As part of the Svelte integration, it equips developers with a compiler that pre-builds the app before the client processes it. Like Vue's SFC format, Svelte uses `.svelte` files, which encapsulate the components with the `<script>`, `<style>`, and HTML tags that are coded together. These are compiled into JavaScript output generated by the compiler.

> **About Vite**
>
> **Vite** (`vitejs.dev`) is framework-agnostic tooling, meaning it can be used in conjunction with different frameworks. It uses a `vite.config.js` configuration file. Mainly, it is used as a build tool for frontend projects. It is optimized for speed and it achieves that speed by providing a development server with Hot Module Replacement and a bundler that optimizes JavaScript output using esbuild (`esbuild.github.io`).

Framework features and patterns

To understand what most modern frameworks enable, we need to understand the following acronyms and features:

- **Single-Page Application (SPA)**: An early term that describes an application that purely uses JavaScript and other frontend frameworks for all interactions with reduced browser routing.

- **Server-Side Rendering (SSR)**: Pre-rendered components on the server side, which are transferred for JavaScript hydration on the client side.

- **Client-Side Rendering (CSR)**: Rendering of components using JavaScript, purely on the browser's side.

- **Static Site Generator (SSG)**: The concept of pre-generating all pages from source for faster rendering and better search engine optimization.

- **Deferred Static Generator (DSG)**: Renders content on the server when initiated by a request to the server.

- **Incremental Static Regeneration (ISR)**: Another pattern of static content generation. In this case, the static generation is triggered by updates by some external trigger.

- **Content Security Policy (CSP)**: Configuration for serving scripts that helps protect against cross-site scripting attacks.

- **Hot Module Replacement (HMR)**: Technique to replace JavaScript modules as the application is running in the browser, mainly used to improve development speed and avoid page reloads.

- **Single-File Component** (**SFC**): A file structure that encapsulates all aspects of a usable framework component, such as styling, templating, logic, and more.

- **Model-View-Controller** (**MVC**): A design pattern focusing on the separation of concerns in various types of applications. It approaches this separation by using the following: a Model that represents the data, a View that provides the user with an interface, and a Controller that is the intermediary between the views and the models.

- **Model-View-ViewModel** (**MVVM**): Another design pattern that also focuses on the separation of concerns in applications, but the approach to these separations is different. In this case, there are still Views and Models, similar to MVC. However, the ViewModel acts as a connection between those types. This approach uses two-way data binding between the View and the Model.

Besides the features and their acronyms, here is a helpful visual describing both the MVC and MVVM patterns:

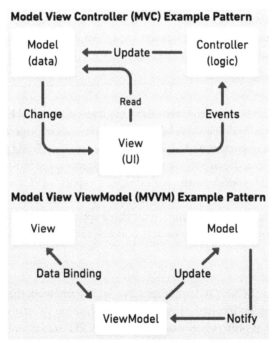

Figure 1.3: MVC versus MVVM patterns

During the renaissance of frontend frameworks, an open source project called **TodoMVC** was established to help developers compare frameworks based on the same To Do app, where anyone can send pull requests with their framework implementations. Besides comparing different frameworks, the project also popularized the approach to complex code organization in JavaScript. Now with the emergence of these new frameworks, we need another iteration of TodoMVC to continue aiding developers with comparisons of these systems.

Backend frameworks

Switching gears from the frontend, let us look at some of the backend frameworks. Node.js plays a vital role in the JavaScript ecosystem, powering a variety of frameworks that allow us to develop backend services. Similar to the frontend, it is impossible to cover all of them, but in this section, we will examine **hapi.js**, **express**, **Sails.js**, **nest.js**, and **AdonisJS**.

Hapi.js

As part of framework explorations over the years, I had the opportunity to work with these frameworks in a professional capacity and on small hobby projects. I started with hapi.js, which is a good example of a well-crafted Node.js framework, built with essential defaults that allow it to craft a server backend quickly. It has a unique approach of avoiding middlewares and relying on external modules. As part of its core, it already has validation rules, parsing, logging, and more built right into it. hapi.js doesn't lock down extensibility; developers can create plugins and register them to execute as part of the different parts of the request lifecycle. Hapi.js' mission puts an emphasis on avoiding unexpected consequences when combining a lot of application logic. This is evident in how hapi.js approaches dependency management and module namespacing.

Express

In stark contrast to hapi.js, the Node.js ecosystem also has a framework called **Express**, which is largely an unopinionated approach to building backend services. Thousands of projects and tools usually use Express for its routing, content parsers, and high-performance reasons. Being flexible in almost every way and with support for over a dozen templating engines, Express is the introductory framework for developers starting with Node.js development. For example, a popular MVC framework, Sails.js, builds upon Express' functionality to offer API generation, database ORM solutions, and support for building real-time features. Generally, it is a good solution for those appreciating the middleware patterns of Express, while having a more structured approach to building backend systems.

NestJS

NestJS, not to be confused with Next.js, is another server-side framework that is worth mentioning. It is similar to Vue, and Angular inspired its approach to application structure, but in this case, for a backend system. By default, it utilizes Express as its default HTTP server and creates an abstraction layer that allows for the ability to change the third-party modules, enabling the developers to swap out Express for some other HTTP framework such as **Fastify**. In NestJS, we see a similar pattern of dependency injection, which enables developers to architect contained modules. These modules can be reused, overridden, and mocked in tests.

AdonisJS

Our final Node.js framework for this section is AdonisJS. Built entirely with TypeScript, it is packed with features that you would expect from a mature framework, such as the ORM based on the Active Record pattern, schema validators, extensive authentication support, and much more. The built-in and first-party plugin features provide solutions for many mundane problems of backend building. AdonisJS also packs a custom templating engine to render HTML layouts. As an added bonus, AdonisJS has straight-to-the-point and clear documentation, which is a joy to read and explore.

Fresh

Given the focus on essential frameworks in the Node.js ecosystem, we should also mention a backend framework called **Fresh**, which is powered by the **Deno** runtime. This runtime is built using a combination of technologies – JavaScript, TypeScript, WebAssembly, and the Rust programming language. Fresh takes a simplistic approach with its emphasis on having no build steps, minimal configuration, and just-in-time rendering of components on the server. Routing is taken care of by creating files in the directories of your project, called File-system routing, a similar pattern in other frameworks.

Looking back at all the Node.js frameworks we covered in this section, there is a healthy framework diversity that delivers solutions for projects of any type.

Native frameworks

The knowledge of JavaScript also allows us to build for native operating system environments and interact with hardware platforms. The availability of the runtime in other environments makes it possible for us to create unique solutions that can help web developers apply their skills in areas beyond the browser. In this section, we cover some of the frameworks created for native JavaScript development.

Electron

The idea of packaging a web app as a native app is not new, but it has been perfected with **Electron**. Electron allows developers to use familiar frontend technologies to build fully capable cross-platform applications that run on popular desktop platforms. It has feature supports features such as auto-updates and inter-process communication, as well as having a collection of plugins that tap into operating system functionality. Besides the advanced framework features, it is beneficial to have a single code base targeting all the platforms, which helps with efficiently building new features and bug fixing. These days millions of people use applications built with Electron, in many cases without knowing it. Applications such as Microsoft Teams, Slack, 1Password, Discord, Figma, Notion, and many more utilize Electron. Even more examples can be found at `electronjs.org/apps`.

React Native

Another framework that helps us create for native platforms is **React Native**, which unlocks the world of mobile development to those experienced with JavaScript. Targeting iOS and Android mobile platforms, just like Electron on desktop, it brings all the benefits of React user-interface building blocks, a unified codebase, and a strong, established community.

Johnny-Five

The Node.js ecosystem also offers hardware frameworks such as **Johnny-Five**, which allows for creative learning use cases of robotics programming using JavaScript and the Firmata protocol. Johnny-Five is an IoT platform supporting over 30 hardware boards. Mainly, it offers interfaces to interact with LEDs, services, motors, switches, and more.

All the frameworks so far deal with building out application logic, but there are also other types of frameworks in JavaScript that play an important role in the development process – these would be the testing frameworks.

Testing frameworks

Testing frameworks in software development are essential for ensuring our projects function as expected. With JavaScript and its supported runtime environments, we have a much more challenging task at hand – we have to test in different browser engines and mock the native web APIs. In some cases, mocking built-in and external libraries can also be challenging. The asynchronous behavior of the language brings its own obstacles as well. Luckily, the JavaScript ecosystem came up with various testing frameworks addressing many software testing challenges – unit, integration, functional, end-to-end testing, and beyond. To name a few, **Jest**, **Playwright**, and **Vitest** all offer great solutions to testing challenges. We will discuss them next.

Jest

As we develop our web applications, we want to ensure that the components we build are functioning as intended; this is where a framework such as Jest comes in. Jest is a unit testing framework that integrates well into other projects. If we are given a project with one of the frameworks that we already saw in this chapter, Jest would equip us with reliable testing solutions. It is ergonomic, with minimal or zero configuration, and provides us with interfaces for easy mocking, object snapshotting, code coverage, and most importantly, an easy-to-understand API for organizing our tests.

Vitest

Vitest is a similar unit testing framework, offering the same interfaces to mock modules in web projects. It focuses on speed and support for components of many frameworks, including Vue, React, Svelte, and even Web Components. It is designed for developer productivity and has a mode for smart test watching, a multi-threaded test runner, and a familiar snapshotting mechanism.

Playwright

Besides unit testing, our software projects benefit highly from end-to-end testing; this is where a testing framework such as Playwright is a good contender. It delivers cross-browser and cross-platform testing for web applications. Playwright comes with a set of test interfaces to automate various browsers, including navigating to URLs and clicking buttons. Historically, this has been a challenging problem due to the asynchronous nature of web pages, but this framework is equipped with ways to avoid flaky tests through retries and await behavior.

Depending on the requirements of the JavaScript project you are involved in, you might have to create new testing workflows or customize the existing testing infrastructure to fit your use case – this is where experience in building testing frameworks would be advantageous.

Framework showcase

Here's a breakdown of the frameworks we covered in this chapter.

These are some of the noteworthy web application frameworks that we will be focusing on in this book:

Frontend + Full Stack		
Name	Released	
AngularJS	2010	Obsolete MVC framework with features such as two-way data binding and dependency injection. Part of the original MEAN software stack.
Bootstrap	2011	Basic framework that allows utilizing HTML, CSS, and JavaScript to create responsive mobile-first websites and can be integrated with other systems to power interfaces for web applications. Bootstrap defines its own layout primitives and provides a great set of built-in components for forms and user interface elements.
Ember.js	2011	Component-service architecture SPA framework with regular releases and opinionated conventions over configuration characteristics.
Vue.js	2014	Lightweight MVVM component-based framework with an easy learning curve – uses the virtual DOM. Comes with its own reactivity system and support for state-changing CSS transitions.
Gatsby	2015	Advanced static site generator using React and Node.js. Includes various modes of rendering pages and serving dynamic websites. Heavily relies on GraphQL for data retrieval. Variety of plugins in the ecosystem.
Angular	2016	Component-based framework with dependency injection, templating, and additive directives. Has a slew of extra features to enable internationalization and accessibility. Full rewrite of the original AngularJS. TypeScript-based.

Next.js	2016	Server-side rendering framework using React as its rendering interface. Supports multiple data request methods. A lot of features are built right into the framework.
Nuxt.js	2016	Framework that uses Vue.js as its core, with a combination of Webpack, Babel.js, and other components. Focuses on delivering an optimized application experience.
SolidStart	2019	Framework for Solid.js applications. Supports all methods of component rendering. Optimizations for code splitting and providing the best Solid.js experience. Solid.js works with real DOM nodes, supports Web Components, and has efficient rendering.
Remix	2021	Full stack, UI-focused framework written in TypeScript. Consists of a browser, server, compiler, and HTTP handler. Built on top of React and includes a powerful application router. Offers many modes of rendering and file-based routing.
SvelteKit	2022	Framework to develop Svelte-based apps. Uses the Svelte compiler and the Vite tooling. Does not rely on the Virtual DOM and supports all modes of rendering components.

Figure 1.4: Examples of frontend and full stack frameworks

These are some of the backend frameworks that will serve as good examples and help us learn certain framework development patterns:

Backend		
Name	Released	
hapi.js	2009	Framework for building backend web services of any kind, with a convention-over-configuration mantra. Supports a lot of advanced features such as API validation, payload parsing, and more right out of the box.
Express	2010	One of the most popular Node.js frameworks for building RESTful APIs, integrated with many modules in the ecosystem. Used in real-world applications and many developer tools. Part of the MEAN stack. Includes helpers for caching, redirection, and support for many templating engines.
Sails.js	2012	Enterprise-grade MVC framework built on top of Express and Socket.io. Comes with ORM support and a powerful CLI to generate parts of projects.
NestJS	2018	Server-side application framework with a modular approach. It follows certain patterns of Angular and includes a lot of built-in features, such as WebSocket, GraphQL, and microservice support.

AdonisJS	2019	All-inclusive backend framework for APIs and web applications written in and for TypeScript-based code bases. Comes with its own components for ORM, templating, and routing.
Fresh	2022	Framework written using the Deno runtime. With no build steps, minimal configuration, and just-in-time rendering. Uses the island architecture pattern, focusing on reducing work on the client. Independent server-side components are rendered using HTML and sent over to the client.

Figure 1.5: Examples of backend frameworks

Other frameworks that use frontend technologies to target native or hardware development are as follows:

Native + Hardware		
Name	Released	
Johnny-Five	2012	Robotics framework for IoT development. Allows developers to interact with hardware modules with an easy-to-use API.
Electron	2013	Popular cross-platform desktop application framework that uses web technologies. Uses the architecture from the Chromium project, which enables developers to interact with the application and the renderer processes.
React Native	2015	Application framework for iOS, Android, and other platforms. Uses familiar concepts from React to build interfaces. Mainly useful for web developers who want a single code base for their application and don't want to use native toolkits to build these apps. Has a large community and plugin ecosystem.

Figure 1.6: Examples of native and hardware frameworks

Here are testing framework examples, which are useful to integrate and use in web application projects:

Testing		
Jest	2019	Zero-configuration testing framework that universally supports many JavaScript environments, including TypeScript, Node.js, and more.
Playwright	2020	End-to-end testing and automation framework supporting cross-platform testing in the Chromium, WebKit, and Firefox browsers. Helps developers quickly and reliably conduct instrument tests for any web application.
Vitest	2022	Unit test framework, part of the Vite ecosystem. Comes with ESM, JSX, and TypeScript support. Developed in conjunction with Vue.js

Figure 1.7: Examples of testing frameworks

One of the best resources to keep up with the current direction of framework development is `stateofjs.com`. It is a yearly survey with results from thousands of developers, and it provides an outlook of where the technologies are shifting. For example, if we look at the frontend framework rankings of 2022 (`2022.stateofjs.com/en-US/libraries/front-end-frameworks`), we can already start to see the retention and interest in React slowly dropping, which could potentially indicate that the industry is slowly shifting to other solutions. Due to this constant change of use, awareness, and popularity of all these frameworks, instead of focusing on many of the frameworks of today, we are going to cover the core patterns that could be applicable to new frameworks in the future. These patterns will be helpful to you while you explore creating your own framework.

Now it's time to try out some of the frameworks mentioned in this chapter using the GitHub repository for this book mentioned in the *Technical requirements* section. You can follow these steps:

1. Install Node.js version 20 from `nodejs.org`.

2. Clone the repository from `https://github.com/PacktPublishing/Building-Your-Own-JavaScript-Framework`.

3. Using your terminal, change directories into the `chapter1` directory of the repository.

4. Run `npm install` and then `npm start`.

5. Follow the interactive prompt to run the examples.

The showcase focuses on reproducing the same example based on the framework type. All frontend frameworks demonstrate the same component written in different structures. Note that some examples might take a while to install and run.

In the final part of this chapter, we shall take a look at my notable personal experiences with frameworks in web development.

My experiences with frameworks

My professional web development career initially started with building basic websites before established frameworks or libraries were around. Looking ahead, I want to share my framework experiences of professionally utilizing them and contributing to some of the open source ones. These days, the accumulated knowledge gathered from these experiences has helped me better assess framework usefulness and introduce new software paradigms into my work. Many find it challenging to keep up with the latest innovations in the JavaScript field, but building even the smallest projects helps one grow as a developer in many ways.

Here are some examples from the full stack development areas that helped me become much more effective as a web developer.

Frontend development

The first few professional websites I built were developed for Internet Explorer 6 and early versions of Firefox web browsers. As we learned from this chapter, there were no frameworks for building web applications at the time, and I had to utilize the few libraries at my disposal. These would help add interactivity for components such as image galleries and dynamic page layouts. Luckily, when my focus switched to larger projects, the **jQuery** frontend library came along and started growing in popularity. Even to this day, jQuery is still a popular tool of choice for a large chunk of websites. I now had the opportunity to hand-craft a basic framework, which could be reused from project to project. This series of scripts was incredibly convenient and foreshadowed the bright future of frameworks that we have today. It was clear that the trend of single-page JavaScript applications was heading towards structured and opinionated solutions.

During one of my large early-on projects – specifically the Firefox Accounts frontend (`accounts.firefox.com`), I had the opportunity to use **Backbone.js**, with the help of jQuery and multiple extension libraries to make it more suitable for large projects. The Firefox Accounts frontend, which is serving millions of users, is still using Backbone.js to this day. The way the Backbone.js framework is structured allows for a soft dependency on jQuery, so it did feel like a natural continuation of my earlier approach to web application development. My key takeaways from this experience are that Backbone.js wasn't the perfect answer to the challenges of frontend web applications, but it was beneficial in many ways. For example, it allowed the project to stay flexible with the ever-evolving JavaScript ecosystem and helped diverse developers work on the application while following a solid set of application guidelines. The unique opportunity to work on the client and the integrated services of the Firefox web browser taught me how to produce JavaScript components for a desktop client that runs on millions of computers worldwide.

Throughout many professional projects, I had the chance to work with Ember.js, Angular, and various React frameworks. I was impressed by how empowering these frameworks can be on these occasions. A notable mention from my experience is the **Meteor** web framework, released in early 2012. One of the big selling features of Meteor was the isomorphic or so-called *universal JavaScript* approach, where the code runs on both the client and the server. In many ways, we see similar approaches in popular frameworks today, where a full stack framework lets developers write JavaScript to develop on both sides of the stack. I have built a few applications and some plugins for this framework, and while it felt so easy to get started with Meteor, I have experienced hurdles while trying to build something that didn't fit exactly into the scope of what Meteor supported, especially in the early releases of the framework. A particular example of fighting with the framework's constraints was developing a feature with a synchronized document state across multiple clients. This ended up being challenging to implement with Meteor's feature set at the time and had to be rebuilt with alternative tooling. Luckily this was not a critical project, but for times when it is important, it is a good idea to evaluate whether the framework of your choice is the right tool for what you are trying to build.

Backend development

During the early years of Node.js, I had the chance to work on several projects utilizing the microservices architecture, and these involved using the Express and hapi frameworks. I felt the contrast between the open-ended approach of the express framework versus the rigorous set of rules and options that were defined in hapi.js. To give some examples, overriding and customizing certain behaviors in hapi.js was quite difficult, and keeping the framework up to date required difficult migrations to the code base.

I still remember combing through the changelog of every new version of hapi.js, making sure not to miss any breaking changes that would make my projects dysfunctional. Even with the hardships of hapi.js, it did feel like the framework was providing a good set of abstractions. In many ways, following the existing examples from something like Flask in Python, hapi had the necessary components to build highly usable services. Meanwhile, my experiences with Express seemed more reminiscent of working with the jQuery and Backbone.js days. In the Express projects, I could have a highly flexible development environment, combining different Node.js modules to achieve what I wanted from the framework. This made me realize that the perfect framework for me would be something between Express and hapi, in the sense that it would allow me to stay creative, highly productive, and able to utilize the runtime ecosystem to the fullest, while at the same time having a strong opinionated framework core, which would keep my application efficient and reliable.

Developer tooling and more

As part of my profession, I have always been passionate about open source, so I focused my efforts on contributing to developer tooling and testing frameworks. I have been a maintainer of **Grunt.js** (`gruntjs.com`), a JavaScript task runner for many years. Grunt.js has been a core component of frameworks, such as **Yeoman**, and has been used as a tool of choice in the early version of AngularJS. The task runner conventions in Node.js have changed a lot since then, but there is still a solid number of projects that use Grunt.js. Maintaining this project for many years feels similar to maintaining a large framework project – releasing new versions, keeping a stable API, supporting it through security bounties, and much more. There is also a huge list of issues, feature demands, pull requests, and plugins to support.

In terms of my testing framework contributions, I was involved in developing the **Intern.js** testing framework (`github.com/theintern`), which enabled unit and functional testing for web applications. I was both the contributor and the consumer of this framework in my daily projects, which gave me a unique angle on the project. I was inspired to provide a good integration experience because it would aid my own projects. As part of this effort, besides learning how a testing framework is built, I focused on developing integration examples and documentation for other application frameworks. Covering many integration scenarios in the provided examples made it much easier for developers to integrate this testing system into their applications.

A final notable framework from my personal experience would be with **voxel.js** – an open source voxel game-building toolkit. While not that popular, it is a great example of creative use of JavaScript, combining both frontend and backend technologies. It is a framework built by a small team that fills in a niche for an audience of developers, who are looking into working on games and visualizations. voxel.js did not set out to be a world-changing framework; instead, it was a creative outlet for many to create. While exploring voxel.js in my personal project, I learned a lot about unique framework and module structures, and it was fun to experiment with systems that enable more imaginative thinking.

Contributing to new projects

These experiences with JavaScript frameworks in frontend, backend, and developer systems were incredibly valuable to me as part of my career. I have learned the importance of following best practices, adhering to software patterns, and developing for various runtime environments, which ultimately helped me write better code and deliver successful projects. As part of this book, I am sharing my learnings and as much knowledge as possible for the next generation of devoted JavaScript developers to build and contribute to their own framework projects.

The projects that I have been involved in always had different origins. In my case, I had to work with both private and open source frameworks. In work projects, I have focused on combining open source tooling with the context of the larger business organization. This approach helped align the existing tooling with the requirements of particular projects. In the open source context, I have been lucky to contribute to projects that have improved the overall developer experience. In many of the scenarios, I got to work on projects that were innovative and were firsts of their kind in the JavaScript ecosystem. For example, when Grunt.js was coming along, there were task runner tools from other languages, but the JavaScript tools were in their inception. Contributing to voxel.js was a similar experience; as more HTML5 APIs and WebGL enable more advanced graphics on the web, it enabled voxel.js as a project and created the contributor community.

During my contributions to the Intern.js testing framework, the overall feeling was that there were not fully fledged testing frameworks that solved all of the needs of web application testing. The goal of this project was to create an all-in solution for testing using the same types of testing APIs.

The framework that we create in this book focuses on the use of modern technologies, such as Web Components, intermixed with popular JavaScript libraries. The Web Components field does not feel as explored in the ecosystem just yet; therefore, with this book, we are taking aim to further widen the knowledge of these technologies among web developers. Besides expanding those skills, one of the greater goals is to make the framework development process more approachable and demystify the established JavaScript systems.

Summary

This first chapter began our exploration into how the web application development process has changed from pure basics into a full-fledged software platform. We have looked at how the innovations and challenges of the web shape the frameworks discussed in this chapter and play a huge role in offering a variety of useful features to web developers. As part of my career journey, working on various projects made me appreciate how much can be achieved by combining elegant patterns and the creative use of the JavaScript programming language.

As part of the framework showcase, it is evident that the ecosystem has a lot of options for the browser and other places where JavaScript runtime functions. However, there is always room for improvement in speed, features, and unique ideas, which can help us enhance the development processes. The significant part that stimulates this ecosystem is the ever-evolving web platform, development of the ECMAScript specification, and of course, the hard work of maintainers of runtimes such as Node.js and Deno.

In the upcoming chapters, we will dive deeper into software paradigms, focusing on framework organization and their architectural patterns. In *Chapter 2*, we are going to look at how frameworks are structured and organized.

2

Framework Organization

The existing JavaScript frameworks have many technical and structural similarities that are useful to learn as part of becoming experts in framework development. In this chapter, framework organization refers to the way of composing a set of abstractions and building blocks, thus creating a collection of usable interfaces that can be utilized in application code.

We are going to learn about the following organizational topics that enable framework development and use:

- Learning about abstractions
- Building blocks of abstractions in JavaScript
- Framework building blocks
- Differentiating modules, libraries, and frameworks

Understanding the core building blocks and aspects of framework development will help us craft our own framework and have the domain knowledge to use other frameworks to their maximum potential. The expectations from the users or stakeholders of a software framework are to have clear instructions with familiar application concepts, reduced complexity, and a well-defined code base. Let us explore how framework organization can help us satisfy those expectations.

Technical requirements

Similar to the previous chapter, we will be using the book's repository as an extension to this chapter. You can find it at `https://github.com/PacktPublishing/Building-Your-Own-JavaScript-Framework`. For the code from the repository, you can use any environment that supports a terminal and runs Node.js, such as Windows, macOS, and most varieties of Linux.

This chapter includes examples from open source frameworks – to save space, the unimportant details are omitted with the `// ...` comment. You will need familiarity with reading JavaScript code, but don't worry if you don't understand the whole code block. As part of reading the code, be sure to follow the links next to those examples to see the full implementation with all the code details. The `chapter2/README.md` file lists the available code resources of the chapter.

In this chapter, it is suggested to try out debugging to further our expertise in framework structures. The easiest way to explore that on your computer would be by downloading the latest version of Visual Studio Code from `code.visualstudio.com`.

Learning about abstractions

Let's dive into the first aspect of framework organization – the fundamental concept of **abstractions**. One of the main conveniences of software development frameworks in web development or other fields is to provide developers with high-quality opinionated abstractions. This means taking a task that can span multiple lines of code, riddled with implementation gotchas, and creating a simple interface around it. It also means coming up with an intelligent way of structuring independent interfaces into a familiar, extensible pattern that is easily usable.

This concept of abstracting away the complexity and generalization of objects helps us define the building blocks that could be used for multiple purposes in our frameworks. Each abstracted object can be initialized with a custom set of properties and be utilized in many shapes when needed. Having this benefit of simplification and generalization in frameworks is what allows developers to focus on the business logic of the programs. Precisely these abstraction concepts benefit developers by eliminating complexity, repetition, and the challenges of learning new systems. Through abstractions, developers do not have to use or even learn the low-level components of the systems they build for.

> **Abstractions in computer science**
>
> In general, the concept of abstractions, a simplified representation of some complex mechanism, is essential in software development. This concept is taught early in programming courses and can be implemented in the high-level and low-level interfaces of large and small programs. Software abstractions structure many parts of these programs and dictate the program's control flow. Data types and structures that define how the data is represented can be considered abstractions on top of lower-level object entities.
>
> Some programming languages offer direct syntax to write abstract classes and interfaces. TypeScript offers this feature as part of its extensions to JavaScript. This allows developers to declare *abstract* classes, methods, and fields. You can find some excellent examples at `www.typescriptlang.org/docs/handbook/2/classes.html` for further reading and prototyping.

If we look at the process of developing websites with bare CSS, HTML, and JavaScript technologies, we can already spot many pre-defined abstractions to make that process more easily accessible and simplified. For instance, HTML, with its elements combined with attributes, can quickly define hyperlinks and embed media with just a few lines of marked-up tags. The styling of those elements with CSS is defined by a set of styling rules targeting particular element nodes. We can see examples of web API abstractions in the **Document Object Model API**, an abstraction on top of a complex tree of nested nodes defining a document structure. These frontend technologies provide a way for user interactions inside a web browser, which generalizes and removes the complexity of interacting with a web application.

We can see a simplified pyramid of abstractions here, starting from application code that developers write on top and rules of basic logic at the lowest level:

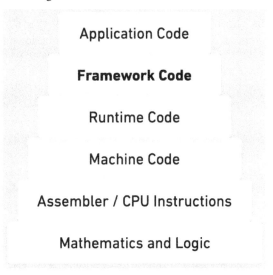

Application Code

Framework Code

Runtime Code

Machine Code

Assembler / CPU Instructions

Mathematics and Logic

Figure 2.1: Abstraction pyramid from higher- to lower-level structures

This finally brings us to JavaScript, as a high-level programming language that already abstracts away complexity in many ways, such as memory management, interaction with the browser, and general object management. In fact, some abstractions that compile to JavaScript focus on abstracting away even the higher-level components. Toolkits such as GWT and programming languages such as Elm, C#, and Dart approach this higher-level abstraction process by compiling to CSS, HTML, and JavaScript. The language extensions for ECMAScript, such as TypeScript, hold closer to JavaScript in terms of syntax, abstract away the common pitfalls in how we write JavaScript programs, and improve the overall developer experience with the addition of a compilation step.

In *Chapter 1*, we looked at several frameworks that use TypeScript, rely on another framework, or both to create a framework of a higher abstraction level. For example, Nuxt.js is a Vue.js framework, and it relies on TypeScript. In such cases, frameworks require developers to use these language extensions and their own self-defined abstractions to build the applications. It is interesting to ponder how deep the levels of abstractions could be nested when we are developing for the web platform and within the JavaScript ecosystem. On the frontend, we have the web browser, which manages the network request/response networking, draws the layout, enables interactivity, and so much more. The backend application services are working on top of the process and operating system infrastructure within the cloud server instances. The levels of abstraction keep growing as we zoom out to electricity flowing through wires, which delivers the required bits for our application code.

Now that we are more familiar with abstractions and why they are used, we shall examine the downsides of this core pattern.

Downsides of abstractions

We have examined the benefits of abstractions, but there are some downsides to consider when utilizing or implementing abstractions. These factors are also true for frameworks and heavily affect framework development. Let's discuss some of the ways abstractions can lead you down the wrong path:

- Abstractions may be incomplete – covering all the potential use cases of an underlying technology with an abstraction could be difficult. For instance, if you have a web application framework, this could usually be a case where a niche feature to output HTML in a certain way could not be supported. The niche requirements could include rendering different types of components, such as SVG animations or direct DOM manipulations. Frameworks offer escape hatches to avoid these issues, but there could be other cases where we have to rely on the knowledge of lower-level components, avoiding the defined abstractions. At the same time, an abstraction may falsely represent the low-level system, which could lead to confusion or false use of the underlying concepts. For instance, if a cryptographic library wrongly uses the primitives, even with the correct outcomes, it could potentially introduce bugs.

- Abstractions introduce an extra layer of code between you and the lower-level system, possibly affecting performance. In the case of frontend development, this means more code to transfer over the wire – additional function calls and layers of indirection. In the backend scenarios, server instances use more process memory. The performance may also be affected by the framework's choice of algorithms. These days, performance is taken seriously by framework authors and users, and regular comparisons and benchmarks help deal with these drawbacks.

- Some of the different framework abstractions may not provide the right interfaces or enough control for the users, which could limit the system's potential. It could be as simple as not supporting all the underlying methods of the lower-level interface. This issue can also happen if a chosen abstraction is used for something other than what it was designed for. This can also be a problem if the framework was designed before a certain technology was introduced. For instance, with **WebAssembly** support in certain frameworks, loading the WASM modules is not possible in some cases due to loading limitations or has to be done with external components. Introducing and using WebAssembly in an already established project with a lacking framework would be an anti-pattern.

Specifically, patterns and abstraction ideas change quickly in the JavaScript ecosystem. New tools and solutions come into existence that abstract how we manage frontend interactions and build backend services. This means, as project owners, we have to adjust either to the changing platform or our existing abstractions become outdated. This could lead to a lack of support for some functionalities or just general code breakage. In many circumstances, this happens when some web APIs change or evolve as the web platform introduces new features.

- Another downside of abstractions is the scenario where a developer may know how to create applications with a certain framework but knows nothing about the internals of the underlying technology. The hidden-away complexity can lead to difficulty in troubleshooting problems and in tracking down errors in the core of the application. Not understanding the technology behind the scenes also limits the developer in their ability to optimize features and take advantage of advanced features.

- We may also face so-called *leaky abstractions*. This is where the attempt to conceal some system complexity fully is not successful. This usually results in the details of the underlying system being revealed to the users of the abstraction. This phenomenon can lead to more complexity in the code with its own problems. The problem becomes evident when developers have to dig into the implementation details of the low-level system and try their best to figure out how the abstraction maps to the underlying system.

- Highly opinionated abstractions, in frameworks and in general, may cause issues when introducing additional layers of sophistication by imposing specific design choices that the developer may disagree with but is unable to change. These can limit the reusability and flexibility of the application code. If we look at Next.js, it provides highly opinionated solutions for several of its features. For example, if you are planning to add API routes to build an API in the Next.js project, these must be mapped to an `/api/` endpoint. To learn more about that, check out `nextjs.org/docs/api-routes/introduction`. This is a simple example, but hopefully it illustrates this drawback well.

No matter where you introduce the use of an abstraction, it does add an extra layer of complexity and indirectness to what we are trying to interface with. When we add abstractions through various means, this makes our projects dependent on them. This dependency may create certain complications. With an external abstraction, we have to accept the risks and trade-offs that come with using it.

In the next section, we are going to look at the popular abstracted building blocks, which are often used in framework development and are exposed as public interfaces of frameworks. We will dive into the frontend browser APIs and the backend runtime modules to better understand what frameworks utilize to build their own abstractions. This is a useful exercise because it helps us figure out how these frameworks function and what techniques they use to combine different tools together. These exercises in tracing the framework organization are valuable to becoming a domain expert in frameworks and understanding the underlying technology behind them.

Building blocks of abstractions in JavaScript

In this section, we will discuss some detailed abstraction examples in JavaScript, as well as web APIs, and features that are used as the building blocks and foundational components of abstractions in frameworks. Framework and browser developers put a lot of thought and hard work into defining these abstractions, which allow developers to be really crafty, produce well-organized code, and build great products.

Frontend framework abstractions

With these three technologies – HTML, CSS, and JavaScript – that enable website development, we get plenty of building blocks that already abstract away the challenges of publishing something on the web. However, we do not get a particular, well-structured, opinionated way to build complex web application projects. This is where the frontend frameworks primarily fill the void that is lacking in the core technologies provided by the web. The frontend frameworks create abstractions in these two cases:

- On top of existing web APIs, which are built into the web browser or a JavaScript runtime.

- When new abstractions are built from scratch based on the framework's internals and opinionated definitions. The innovative and unique approach to these abstractions is what makes a particular framework desirable and liked among developers.

The following web APIs provided by the browser engines are often abstracted by frontend frameworks:

- **Document Object Model** (**DOM**) – This allows manipulation of the web page's structure. The DOM represents a tree where the nodes constrain the objects. The DOM API provides access and the ability to modify this logical tree. The user interface frameworks primarily need this to display the rendered views and handle DOM interactions and events. Even the frameworks that use a virtual DOM need to attach their structures to the real document to make it visible on the page.

- **Browser Object Model** (**BOM**) – This allows manipulation of browser-specific properties such as browser history, location, screen, frames, navigator, and more. Technically, this also has access to the DOM's `document` property. The browser also provides complex APIs, such as the WebAssembly API, which lets applications include binary-code modules. The frameworks usually have a loader for these low-level modules as part of their loader workflow.

- **CSS Object Model** (**CSSOM**) – A set of browser APIs to dynamically manipulate, read, and modify page style. Similar to the DOM APIs, frameworks can utilize CSSOM for custom solutions for styling and animation behaviors. The capabilities of CSSOM allow JavaScript to control element styles and more. On a basic level, it offers the following programmatic access: `document.body.style.color = 'pink';`. This object model also offers several method calls such as `getComputedStyle();` to fetch information about the object's style.

- **Network APIs** – These APIs have the ability to make asynchronous network requests using the Fetch API or the XMLHttpRequest API. Frameworks utilize these for basic networking operations, including structuring complex requests using GraphQL. The networking APIs also offer **WebSocket** functionality. These APIs provide full-duplex (data can be transmitted and received simultaneously) communications with less overhead than usual networking calls, enabling applications with real-time updates and communication. The WebSocket API is simple enough that it can be used directly in the applications or with the inclusion of an extension package for a particular framework. **Socket.io** builds on top of WebSocket APIs and offers a complete low-latency solution that can coexist with the framework code. Finally, **WebRTC** also fits into the category of networking APIs, and it enables the capturing and streaming of audio and video content in the browser. Similar to WebSockets, the WebRTC framework integration is usually included with an external library because it is quite a subtle feature.

- **Storage APIs** – These APIs have the ability to store data for web application and caching purposes. These APIs are often utilized to store data in the local and session storages. They also write to browser cookies and databases, such as **IndexedDB**. For instance, Angular applications can include a dependency that provides a cookie service and makes it easier to read and write cookie information.

- **Background services** – These include a slew of services that enable background synchronization of data, notifications, push messaging, and more. Web workers generally provide a way to run background-independent scripts and make use of multiple CPU cores.

- **Graphics APIs** – These APIs grant the ability to render high-performance 3D and vector graphics. These include the **WebGL** APIs and **SVG** elements. The 3D graphics libraries utilize the *canvas* element for rendering and can utilize the graphics hardware. For applications built with Vue.js, there is an additional component library called VueGL that makes it easier to create WebGL-based components. In terms of SVG, JSX in React is able to directly parse SVG syntax, as long as the SVG properties are converted to the camel-case JavaScript syntax.

A useful, in-depth list of web APIs that frameworks can potentially utilize can be found at `developer. mozilla.org/docs/Web/API`.

Let's now take a look at a real-world frontend example with Nuxt.js.

Real-world examples

Nuxt.js uses Vue as a backbone for its frontend components. As shown in *Figure 2.2*, the Nuxt.js framework has the `NuxtLink` built-in component for creating links, which can be used in application code, and utilizes several modules from Vue, such as the `vue-router` and component-building functions such as `defineComponent`:

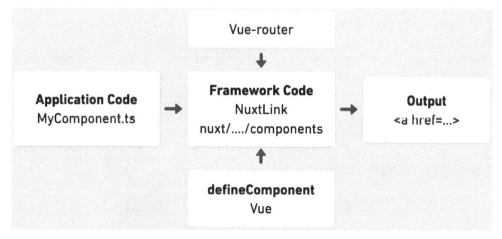

Figure 2.2: Nuxt.js and Vue framework abstraction

We can follow the usage of this abstraction (*Figure 2.2*) in detail from an `example.vue` page in application code:

```
pages/index.vue
<template>
  <NuxtLink to="/">Index page</NuxtLink>
  <NuxtLink href="https://www.packtpub.com/" target="_blank"">Packt</
NuxtLink>
</template>
```

To run this particular code example on your computer, navigate to the `chapter2/nuxt-js-application` directory and run `npm install && npm run dev`. Refer to the included `chapter2/README.md` documentation for additional details. Once the application is ready to run locally, you should see the URL that you can open in your browser in the terminal. Here's an example of the successful output:

```
> npm run dev
  > Local:    http://localhost:3000/
  > Network:  http://192.168.1.206:3000/
```

The auto-import feature of the framework allows the direct use of the component to create the two links in the template file. This built-in component is defined in the source of the Nuxt framework at `github.com/nuxt/nuxt/blob/main/packages/nuxt/src/app/components/nuxt-link.ts`.

Let us take a moment to understand the `nuxt-link` component code, following along with the source from the framework itself. The code behind this particular component extends the routing behavior from Vue.js. It defines the typed TypeScript interfaces, such as `NuxtLinkOptions` and `NuxtLinkProps`, to accept particular styling attributes and routing options. The `defineNuxtLink` function returns the component with the customized routing behavior. Helper functions, such as `checkPropConflicts` and `resolveTrailingSlashBehavior`, address specific routing use cases. The `setup()` function call uses the Vue 3 Composition API to enable reactive properties of the component and attach component lifecycle hooks to versions of `NuxtLink` in applications. More details on this API can be found in Vue.js' documentation – `vuejs.org/guide/extras/composition-api-faq.html`. The important parts are shown in a condensed form here:

```
export type NuxtLinkProps = {
  to?: string | RouteLocationRaw
  href?: string | RouteLocationRaw
  target?: '_blank' | '_parent' | '_self' | '_top' | (
  string & {}) | null
  // ...
}
export function defineNuxtLink (options: NuxtLinkOptions) {
  const componentName = options.componentName || 'NuxtLink'
```

```
  return defineComponent({...})
}
```

In the preceding component code, we see the final return statement of `defineComponent`. This generates the anchor `<a>` element that we ultimately see in the final source of the HTML structure in the web application. It is produced by the internal Vue.js function call:

```
  return h('a', { ref: el, href, rel, target})
```

From the `defineNuxtLink` function, it is also evident that it is possible to modify some parts of the component. For example, we can define a component with a custom name using the `componentName` parameter.

From *Chapter 1*, we've seen that JavaScript has much to offer for full stack frameworks. In the next section, we are going to look at the APIs that we can utilize as part of the backend environment.

Backend runtime abstractions

Although there are many frontend APIs to choose from, we are still bound by what the web browser supports. This is a lesser issue when we write backend services, as there are still many APIs that we can use, and we can even write our own extensions or integrate with external systems for custom use cases. Let's take a look at important APIs that we can use as part of framework development.

In this section, we will look at both **Node.js** and **Deno**, as they are two runtimes that offer similar features. These runtimes need to handle server creation, file and process management, module packaging, and more. The following are some of the essential APIs that are used by the backend frameworks:

- **Filesystem APIs** – This has the ability to read and write files and other file system entities. Frameworks use this heavily while storing data, loading existing files, and serving static content. These APIs also include file streaming and asynchronous features.

- **Networking** – These are the APIs to start new server processes and accept requests. Includes handling of HTTP requests and other request formats.

- **Modules and packaging** – These are the conventions of how modules and packages can be loaded.

- **Operating system APIs** – These are the APIs to fetch information from the operating system that is running the process. This includes useful data about memory consumption and useful operating system directories.

- **Process handling** – These APIs allow for manipulating and gathering details of the currently running process. These also enable sub-process creation and handling of multi-process communication.

- **Native modules** – The native module APIs allow users to call out to libraries written in other native languages, such as C/C++, Rust, and others. In some cases, they use the **Foreign Function Interface** (**FFI**). WebAssembly is also part of this native module support.

- **Worker APIs** – Allows the spawning of additional worker threads to schedule heavy server work to happen outside of the main process. For instance, the Deno runtime supports the **Web Worker API** to provide these features, while Node.js uses its **Worker Threads** module.

- **Console and debugging** – This set of internal APIs allows for recording process logs. The debugging APIs make developing and finding issues in the running code easier. Paired with an editor that supports debugging operations, it can pause the debugger when the framework request handler is processing a request.

These are some of the APIs that backend frameworks can use as the foundation for their projects. For instance, the hapi.js framework is able to combine some of these APIs to create its `Server`, `Route`, `Request`, and `Plugin` modules. For example, its `Core` (`hapi/lib/core.js`) module makes use of the operating system, networking, and module-handling APIs.

Next, we can take a look at a detailed example of combinations of abstractions and runtime APIs in Nest.js, a framework that is familiar to us from *Chapter 1*.

Backend framework abstractions

The Nest.js framework supports the ability to provide any HTTP framework as long as there is an adapter defined to work with it. The existing adapters that are built right into Nest.js are `platform-express` and `platform-fastify`. The default behavior of the HTTP adapter abstraction is transparent to the developer as it uses the `express` module by default.

In *Figure 2.3*, we can see the combination of all the components. The application code is powered by the framework that utilizes both the framework abstractions and the Node.js API:

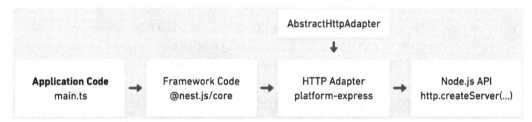

Figure 2.3: Nest.js framework abstractions

As shown in *Figure 2.3*, the `main.ts` entry point file in Nest.js starts the server and listens for incoming requests:

```
main.ts
const port = 5300;
const app = await NestFactory.create(AppModule);
await app.listen(port);
```

express-adapter, which extends from AbstractHttpAdapter, defines the set of methods required for the HTTP server, including the .listen method:

```
import * as express from 'express';
import * as http from 'http';
import * as https from 'https';
// ...
export class ExpressAdapter extends AbstractHttpAdapter {
  // ...
  public listen(port: string | number, callback?: () =>
    void): Server;
  public listen(
    port: string | number,
    hostname: string,
    callback?: () => void,
  ): Server;
  public listen(port: any, ...args: any[]): Server {
    return this.httpServer.listen(port, ...args);
  }
  // ...
}
```

The adapter code above utilizes the express framework and the internal http APIs. Ultimately, this results in a class that exposes a method to set up an HTTP server. Even though express provides the routing and HTTP helpers, it does not start the server by itself. Inside express-adapter, there is a direct call to Node.js APIs:

```
initHttpServer(options) {
  const isHttpsEnabled = options && options.httpsOptions;
  if (isHttpsEnabled) {
  this.httpServer = https.createServer
    (options.httpsOptions, this.getInstance());
  } else {
  this.httpServer = http.createServer(this.getInstance());
  }
  // ...
}
```

The direct call in the preceding code block figures out the type of server to start, *HTTP* or *HTTPS*. It also accepts a variety of httpOptions values. This pattern is similar in other frameworks. For instance, in AdonisJS, the framework authors define the HttpServer class (github.com/adonisjs/core/blob/master/src/Ignitor/HttpServer/index.ts) that creates an HTTP server and utilizes the createHttpServer utility function to call out to the runtime APIs of Node.js.

As we dig in further to understand how existing frameworks are structured and how their abstractions work, it is important to have a way of traversing the code of these nested abstractions. In the next section, we will cover the technique of debugging, which can help us uncover the hidden interfaces within the frameworks.

About debugging

Debugging plays an important role in software development. It helps us identify and resolve issues quickly. As part of the framework learning process, it also helps us understand how these frameworks work internally. By stepping through the breakpoints of the program and digging deep into the call stack, we can understand the inner workings of the internal modules. It also helps us navigate through multiple levels of abstractions.

Node.js' debugger integration provides a way for us to debug our programs and frameworks. It is a good habit to try it out on your own to get a better understanding of how the framework functions. For example, to debug a Nest.js application, we can utilize the Visual Studio Code debugger:

1. Open up the `nest-js-application` project in the `framework-organization` directory of the book's GitHub repository.

2. Run `npm install` to get the project's dependencies.

3. Set a code execution breakpoint in the `app.service.ts` file of the application; refer to the screenshot in *Figure 2.5*. To set the breakpoint, hit the empty space to the left of the line number until you see a red dot. Once this red dot appears, that will be your breakpoint.

4. In Visual Studio Code, browse to the `package.json` file, and press the **Debug** button near the `scripts` section. See an example of this in *Figure 2.4*:

```
      ▷ Debug
 8      "scripts": {
 9        "build": "nest build",
10        "format": "prettier --write \"src/>
11        "start": "nest start",
12        "start:dev": "nest start --watch",
```

Figure 2.4: Debug button in package.json

5. After pressing the **Debug** button, you will be presented with a context menu of all the available scripts in the project. Select the `start:dev` option, which should start the application, observable in the **Terminal** tab of Visual Studio Code. Watch out for errors in the terminal log. For example, if the `nest` command is not found, that means you need to install the dependencies for this project using `npm install`.

6. With the application running in debug mode and using the `start:dev` script, open the address at `http://127.0.0.1:3000` in your browser. This should now pause on the extract line of your breakpoint.

If you hit a breakpoint properly in the editor, that means you successfully attached the debugger to the application. You can now use the call stack pane on the left (as shown in *Figure 2.5*) to navigate around the running process and browse through the Nest.js modules:

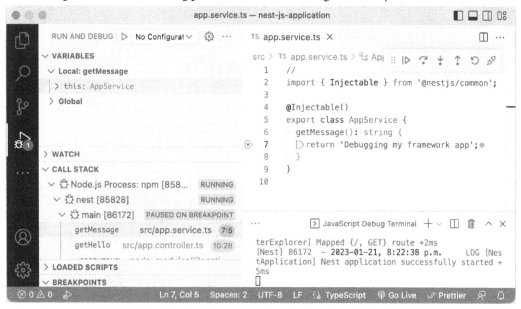

Figure 2.5: Debugging a Nest.js application

This technique is a quick way to get a view behind the scenes of a framework. It enables developers to quickly get a sense of how the framework functions and makes it easier to understand the nested abstractions. To get more from it, you can find an in-depth explanation of the Visual Studio Code debugger at `code.visualstudio.com/docs/editor/debugging`.

Framework building blocks

Just like most programming languages, JavaScript, and its extensions such as TypeScript, have the fundamental features of working with numbers, strings, Booleans, conditional logic statements, and much more. The more advanced features are built on top of those fundamentals. The frameworks utilize existing interfaces, such as events and modules. However, they also create their own build blocks, by defining interfaces to create components, routers, and others.

In this section, we are going to examine both existing interfaces and custom ones. We are going to look at some of the common interfaces, which can be combined to make a framework. These are the abstracted entities that solve particular problems in application development and are beneficial to their users.

Events

Event binding and events are everywhere in JavaScript applications. They enable frontend user interfaces and interactivity through buttons, forms, pointer movement, keyboard keys, scrolling, and more. The concept of event binding is something that every framework handles with different syntax definitions as follows:

```
// Vanilla JavaScript
someInput.addEventListener('keyup', keyDownHandler)
// Vanilla HTML
<input type="text" onkeyup="keyDownHandler()" />
// React
<input type="text" value={answer} onKeyPress=
  {keyDownHandler}/>
// Angular key down combination of SHIFT + ESC keys
<input (keyup.shift.esc)="keyDownHandler($event)" />
// Vue key down
<input @keyup.shift.esc="keyDownHandler" />
```

Most of the time, the event handling is very similar to original DOM events, but with a modified syntax to fit the framework abstractions better. Frameworks further enable event handling by providing more complex event management components. For example, Angular has the concept of `HostListener` (`angular.io/api/core/HostListener`) to register events within its components.

On the server side, Node.js is heavily event-based because of its asynchronous, event-driven architecture and the frameworks take advantage of that. For example, hapi.js maintains its own event emitter package called `@hapi/podium`, which allows the developers to register custom application events.

Another example of the event handling pattern would be how the **Johnny-Five** framework handles sensor data. If the hardware board has a GPS chip, then it will use `change` and `navigation` events to report the latest values:

```
board.on("ready", function() {
  const gps = new five.GPS({
    pins: {rx: 11, tx: 10}
  });
  gps.on("change", function() {
    console.log(this.altitude);
  });
  gps.on("navigation", function() {
    console.log(this.speed);
  });
});
```

The use of events is an important building block, which allows us to subscribe to user interaction and listen to the changes or progress of some operation. As we are set on developing our own framework, it needs to provide a way to interact with events and abstract certain complexities around them.

Components

Many frameworks provide an abstraction to create reusable components to organize the project. Depending on how the application is planned out, components can help split any type of application into reusable and independent pieces of code. These code parts can also be nested within each other. Depending on the required business logic, developers can define custom components, use the pre-built ones, or import a library of components that are designed for specific use. Once many components are nested and situated together, it is common to have some interaction between these objects. The components utilize data properties to render the information to the user from the current state, and in many cases, they need some properties from the parent components. For frameworks that use React or Vue, this means writing communication patterns that enable child-to-parent component communication and the other way around. This communication process can get complicated, which is why these frameworks use a unidirectional or a one-way data flow where the data updates flow from parent to child components. Instead of synchronizing the same state between nested components, it is advised to store the state in the most common ancestor component in the chain.

If we have a complex application, this means that we will probably end up with a lot of components, nested multiple levels deep. This is where **component composition** can help. Component composition is a pattern that allows for minimal code duplication, and performance improvements.

In the following figure, we have an illustrative example of how composition can affect and reorganize a set of nested components within an application. The component organization pattern is very familiar to developers, so it would be a good choice to use or create a framework utilizing this pattern:

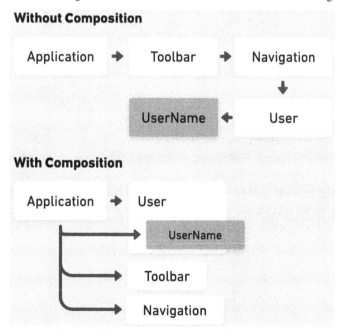

Figure 2.6: Nested versus composed components

Lifecycle methods

Lifecycle methods or events are often managed by the framework, providing the ability to execute code at particular points. These methods can be used to execute custom logic at different stages of components and other parts of the system, which provides flexibility to the framework interfaces. These lifecycle methods can be used to attach or detach additional logging, utility functions, and more during the component execution. Lifecycle sequence, which means the order of how these events occur, must be well documented and described in the framework. This is done mostly due to the fact that the lifecycle methods can have a particular naming convention and have a complex runtime hierarchy.

In Nest.js, the server framework provides lifecycle hooks to its module system. Some examples of that are onApplicationBootstrap(), which is called when all modules in the application have been initialized, and onModuleInit(), which is called when the dependencies of a module have been resolved. Using the TypeScript interfaces in Nest.js, we can inject code into the onApplicationShutdown lifecycle event as all connections to the server close down, which can be defined as follows:

```
@Injectable()
class SomeService implements OnApplicationShutdown {
  onApplicationShutdown(processSignal: string) {
    // ...
  }
}
```

In Vue.js, given that the framework deals with rendering components, the events available cover the whole lifetime of the component. For instance, it has beforeCreate, created events as the component initializes its state, and beforeMount, mounted events as the component gets mounted to the DOM tree. You can find a great lifecycle diagram of Vue.js at vuejs.org/guide/essentials/lifecycle.html#lifecycle-diagram.

Router

Both frontend and backend frameworks usually require some form of a router that navigates to different parts of the application. The router mechanism on the frontend follows the navigational pattern of web pages, following the browser's URL patterns. On the frontend, the router is essential to transition between states or navigate to internal or external pages. Besides providing the routing *tree* structure, the router is also responsible for interfaces that allow components to invoke the routing behaviors – we saw an example of this via NuxtLink in the *Real-world Examples* section of this chapter.

The **react-router** (`reactrouter.com/main/start/overview`) project is a good example of everything you need from a router component. It makes it possible to simply define the routes within a component as follows:

```
<Routes>
  <Route path="/" element={<Layout />}>
    <Route index element={<Login />} />
    <Route path="register" element={<Register />} />
  </Route>
</Routes>
```

The backend server frameworks use a server router to handle requests coming to API endpoints. Normally, the router interface takes different forms of endpoint URL structures and maps those into functions that process the route. Some good examples of an unopinionated router can be found in Express.js (`expressjs.com/guide/routing.html`).

From the following code example, we see the relationship between the endpoint path and the function that is able to process the request and a text response:

```
app.get('/framework-organization', (req, res) => {
  res.send('Learn about framework organization!')
})
```

From *Chapter 1*, we have seen examples of file-based routing, which further simplifies the routing mechanisms by just looking at the files in the application, and dynamically creating routes based on those files.

Template engine

Another essential building block is the template engine. This engine combines the static parts of a marked-up document with the data from the application. Templating makes it possible to render views with various forms of data. With frontend frameworks, this usually means rendering the nested component hierarchy. The template engine's job is to enable data binding and to bind any specified events for the interactive components, such as buttons or input fields.

With a backend framework, templating engines render the whole page or, in some cases, partials, sending them over the wire to the client for initial static rendering. From *Chapter 1*, we have seen frameworks, such as Next.js, that are able to render the frontend components on the server side and then attach any JavaScript behaviors to the readily rendered component. By default, in Next.js, the pages are pre-rendered to improve search engine optimization and performance in the browser client. Templating is a vital building block of a framework – this is how developers create the presentation layer and mark up the page structure.

Networking

Web frameworks commonly provide several components as networking abstractions. The ability to use a good networking interface can vastly improve the readability, performance, and error handling of the application. Here are some features that usually come as part of a good networking abstraction:

- **Session management** – This is the ability to manage sessions and provide easy access to session information. This is included as part of networking because frontend frameworks usually rely on the backend service to fetch and parse the session information.

- **Error handling** – This provides good interfaces to handle all types of possible errors that can happen during the process of making a request to an endpoint.

- **Caching** – This is the mechanism that provides a caching layer to improve performance and avoid redundant queries if the data is already fresh enough.

- **Security** – Often frameworks come with baked-in security features that follow the best practices. This includes examples such as XSS, CSRF protection, script injection prevention, and input validation.

- **Request and response management** – This improves the ability to make requests with the required parameters and parse the responses from external systems.

Most of these networking abstractions apply to both frontend and backend systems. In a full stack framework, a combination of these abstractions can vastly improve the workflow and the efficiency of the system that it is supporting.

All these abstractions are implemented in JavaScript, TypeScript, or enabled by the runtime. Their implementation can be provided from within three categories of code structures – as a module, library, or built into a framework. In the next section, we are going to explore these categories of code organization.

Differentiating modules, libraries, and frameworks

While working on JavaScript applications, we rely on modules, libraries, and, of course, the larger frameworks. These structures can originate from internal and external sources, meaning they are either written by you or your team or are a dependency that is written by someone else. JavaScript, specifically, is in a unique position where modules, libraries, and even frameworks can be used on browser and server environments. For framework developers, it is important to know how to work with these JavaScript structures, because frameworks heavily rely on defining and using modules and libraries. These abstractions and structures allow for better code organization, which we will be discussing in the next subsections.

Modules

Developers create their own modules to separate the code into multiple files or logical blocks. In a similar manner, modules can be imported from external sources. The module encapsulation wraps a block of code, providing a way for it to export various data types through strings, functions, and other data types.

The history of how modules are defined and used in JavaScript is complicated; it ventures out to different module patterns and implementations. Even in modern projects, you will find inconsistent approaches to module management. Initially, there was no way to organize modules, so frontend JavaScript modules were wrapped in immediately invoked function expressions or objects. Using functions allows the hosting of all the values inside of it within a lexical scope. Here is a sample:

```
let myModule = (function () {
  // ...
  return {
    someProperty: function ()  {
    // ...
    }
  }
}) () ;
```

It was evident that the language needed some kind of a modular pattern, and this is where **CommonJS** and AMD types of modules were introduced. Defining a CommonJS module is straightforward and can be seen in use in a lot of the Node.js frameworks:

```
module.exports = class MyModule {
  constructor(someProperty) {
    this.someProperty = someProperty;
  }

  myMethod() {
    return this.someProperty;
  }
};
```

For instance, hapi.js uses a similar CommonJS module pattern in most of its files, which can be found in the framework repository at github.com/hapijs/hapi/tree/master/lib – a simple example from a list of its modules is lib/compression.js in the lib directory:

```
const Zlib = require('zlib');
const Accept = require('@hapi/accept');
exports = module.exports = internals.Compression = class {
    // ...
    accept(request) {
        const header = request.headers['accept-encoding'];
        // ...
    }
};
```

This module provides compression functions for many hapi.js use cases. With some of the code omitted, we see the exports keyword, which is used to make methods from this module available in other files.

These days projects may have different types of JavaScript modules that are used as part of their workflow. The more standard ones you will see are CommonJS and **ECMAScript Modules** (**ESM**) module files. We saw an example of CommonJS structures earlier in this chapter with its `module.exports` keywords. The ESM system provides `import` and `export` keywords to manage the modules. To distinguish the module types, the `.cjs` and `.mjs` file extensions are used to be explicit about which module system is used. The normal `.js` extension can still be used, but then it is up to the module loading system to figure out how to load these files.

Some examples of module types that you may see in the ecosystem are listed here:

- **Universal Module Definition** (**UMD**) – This is the module definition that tries to support all possible module declarations

- **Asynchronous Module Definition** (**AMD**) – This is a largely deprecated system with `require()` and `define()` functions to manage the module

- **Immediately Invoked Function Expression** (**IIFE**) – These are simplistic modules encapsulated by a function scope

The module systems are slowly getting better in JavaScript, but it is something to watch out for when using various modules in frameworks and choosing the right module system to use. The varying usage of modules can lead to issues with loading in specific environments or can lead to some features not working as intended.

Libraries

Libraries are indispensable in software development nowadays; they play an important role in supporting web development projects of any size. The libraries consist of a collection of useful resources that implement certain specific functionality with a well-defined interface. The focus of libraries is to include the encapsulation of optimized features to solve certain problems. Most libraries try to focus on a specific set of problems that can help their stakeholders. JavaScript has an abundance of open source libraries that developers cannot live without in professional projects. They don't dictate any specific opinionated control flow, but instead, let the developer make use of them when needed. In a similar fashion to frameworks, the technological availability of the JavaScript runtime allows some of the libraries to be used in both the browser and the backend environments.

Libraries can serve as a core component used by a framework to solve certain technological challenges. There are many cases where we see frameworks building abstractions around the following libraries:

- **lodash** (`lodash.com`) – This library provides a large set of utility functions for common tasks

- **React** (`reactjs.org`) – This is an extremely popular user interface component rendering library with state management that many of the frontend frameworks mentioned so far are based on

- **Axios** (`axios-http.com`) – This is a powerful HTTP client library for frontend and backend JavaScript projects

- **Luxon** (`moment.github.io/luxon`) – This is a library to manipulate date and time in JavaScript, the evolution of a popular library called moment.js

- **jQuery** (`jquery.com`) – This is more than a decade old – a popular library that simplified DOM traversal and abstracted away cross-browser quirks for CSS, AJAX, and more

- **Three.js** (`threejs.org`) – This is a JavaScript 3D library that abstracts away the complexities of WebGL and 3D graphics on the web

It is a common pattern to build out additional tooling on top of existing libraries to enable applications of various types. For example, in *Chapter 1*, we saw examples of frameworks building tooling and abstractions around React. If the project allows you to, it is usually a good idea to utilize a library or learn from existing implementations for a solved problem rather than rebuilding or rewriting the same code. In the next section, we are going to compare choosing a development workflow with a set of libraries in contrast with framework-driven workflows.

Frameworks

From *Chapter 1*, we already know what JavaScript frameworks do and the benefits they offer. It is also important to understand how much frameworks rely on libraries for their workflows and features, and the differences between them. Both libraries and frameworks manage control flow in the applications. This control flow is the order and structure of how the logic in the application flows. With a library workflow, the existing program will have its own control flow and as part of it the library functions are executed when required. This existing program precisely calls out to the reusable code in the library. This gives full control to the developer to structure the application as they see fit, allowing for more room to fine-tune the application behavior but missing the value and structures that could potentially be gained by using a framework.

With a framework workflow, the framework dictates how the control flow should be structured. In this case, the developer works within the constraints of the framework and follows the often strict guidelines defined by someone else.

A good comparison would be the React library versus a JavaScript framework such as Next.js that depends on the library. The library just consists of functions that execute certain tasks. React contains logic for rendering, creating components, and other methods. But it is Next.js – the framework – that defines the architecture for the application, using the library methods from within to enable its features.

Choosing the workflow of using the library within a framework makes it a powerful combination; this way, it is possible to gain the benefits of both of these tools.

Framework organization showcase

You can try out the framework organization examples from this chapter in the book repository. Access the examples by cloning the repository from `https://github.com/PacktPublishing/Building-Your-Own-JavaScript-Framework`. Then use your terminal to change directories into the `framework-organization` directory of the repository and run `npm install` followed by `npm start`. Follow the guidance in the terminal and keep an eye out for the `README.md` files in the directories for extra information.

Summary

We have looked at abstractions combined with clever API design, which together form the key to building successful frameworks. We have also expanded our knowledge of common framework interfaces that provide value to developers and make the application development process much more efficient and approachable. Understanding how modules, libraries, and frameworks are used helps us be better system architects. In addition, the ability to use the debugger to quickly explore how all these pieces come together in existing frameworks empowers us to be much more efficient developers.

Now that we are more familiar with various organization patterns, we can dive deeper into the specific techniques that help us architect new systems. In the next chapter, we are going to look at the existing and common patterns that combine the building blocks that we have learned into a cohesive system.

3

Internal Framework Architecture

In the previous chapters, we learned about the history of the current frameworks and explored the concept of abstractions. We also looked at how JavaScript frameworks use, combine, and extend different abstractions to make a framework functional. In this chapter, we will dive into the architectural patterns of JavaScript frameworks. To go even further and expand our framework knowledge, we will have to closely inspect what goes into making a modern framework and what we can learn from existing patterns.

As part of this chapter, we will explore the following:

- Understanding the core technical architectural patterns of existing frontend and backend frameworks. We will focus on the design, architecture, and structural patterns that are combined into a single system.

- Getting a glimpse into framework APIs, packaging configurations, and additional tooling.

- Understanding additional tools that benefit the framework.

- Exposure to available compiles and bundlers.

Technical requirements

Similar to the previous chapter, we will be using this book's repository as an extension of this chapter. You can find it at https://github.com/PacktPublishing/Building-Your-Own-JavaScript-Framework, and the relevant files are in the chapter3 directory. To run the code in this repository, you can use any environment that supports using a Terminal or a command prompt, running Node.js, such as Windows, macOS, and most varieties of Linux.

Most of the code mentioned in this chapter can be found in this book's repository, so you don't have to navigate external repositories. Follow the README.md instructions in the chapter3 directory to get started. In addition, a lot of framework architecture focuses on the stakeholders or developers who will use the framework to build new projects. In this chapter, we will refer to them as *framework users*, not to be confused with the end users of applications built with these frameworks.

Exploring the core architecture

Let's explore the core architectural pieces of building a framework. When a framework project is created, it is usually partitioned into an organized directory structure consisting of various specialized compartments. This approach helps separate the concerns of specific modules, scripts, and files. This organizational pattern is similar to how web application projects are organized. Except in the case of the framework, the project needs to export public interfaces and scripts to be usable. In some cases, frameworks can also be split into multiple repositories to allow a different approach to framework development.

Given that the programming environment for every language is different, the JavaScript and TypeScript frameworks have their own ways of structuring the framework projects, making it easier to generate artifacts and make the frameworks more usable within the projects that utilize them. A well-organized project makes it easier to maintain, collaborate, and refactor many parts of said project. For instance, let's take real-world examples such as Vue.js and Angular. Angular keeps most of the framework files in a single repository (which can be found at github.com/angular/angular), except its command-line tools (located at github.com/angular/angular-cli). However, many packages are published as separate dependencies in the package registry, such as **npm** (which can be found at npmjs.com/package/@angular/core). The resulting payload for the framework package can be different from a frontend framework. In the case of Angular, it consists of pre-built files for different versions of JavaScript.

In *Figure 3.1*, Angular core is provided as unflattened and flattened versions in the esm2020, fesm2015, fesm2020 directories, and these packages also include various source maps:

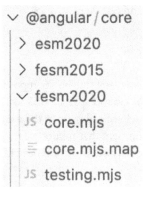

Figure 3.1: Angular core files provided as an npm package

Depending on the browser target or the bundler technology used, Angular offers multiple import options. Vue.js has a similarly packaged export that offers a myriad of options to load the framework in a particular JavaScript module environment.

The output directory in *Figure 3.2* is the result of packaging the Vue.js framework using `rollup.js` (`rollupjs.org`) to create an output configuration for each of the target runtimes. The source of this configuration can be found at `github.com/vuejs/core/blob/main/rollup.config.js`:

Figure 3.2: Vue.js distribution as an NPM package

The produced files of the targeted configurations slowly evolve with the runtime target as time progresses. For instance, if the framework is not usually included as a global variable inside of a `<script>` tag, then it might make sense to get rid of the global variable output or let the framework users convert the resulting output of the framework to suit their needs.

Examples of packaged frameworks

You can check out the examples of the framework bundles by running the `npm start` script in the `chapter3` directory. Once the packaged framework sources have been downloaded and extracted, you can view several resulting outputs for Angular and Vue.js.

Now that a build step is so widespread in JavaScript projects, these frameworks also provide direct ways to import the framework packages and include them as part of the build process. The "unflattened" version of the framework would be an example of this, where the framework files are not concatenated. Instead, the bundler would be combining those files and optimizing them with techniques such as code splitting if possible.

The applications that use the framework import the dependencies in the `package.json` file. Angular splits the different parts of the framework into different packages:

```
"dependencies": {
    "@angular/animations": "^15.2.0",
    "@angular/common": "^15.2.0",
    "@angular/compiler": "^15.2.0",
    "@angular/core": "^15.2.0",
    "@angular/forms": "^15.2.0",
    "@angular/platform-browser": "^15.2.0",
    "@angular/router": "^15.2.0",
},
```

Splitting these modules into their own packages creates a well-defined boundary between the different modules. However, extra release tooling is usually required to make it easier to manage and release multiple modules without manually packaging them. The split packages also benefit framework users as they can pick and choose which modules are required for their applications.

> **Angular core dependencies in action**
>
> The framework showcase from *Chapter 1* has an existing example of all the Angular dependencies mentioned in this section. You can find the application in the `chapter1/angular` directory, run it, and tweak it to your preference. Many of these core dependencies can be found via the `@angular` namespace in the npm package list at `npmjs.com/search?q=%40angular`.

In the case of Vue.js, the core architecture is divided across many repositories at `github.com/vuejs`. The core, router, developer tools, documentation, and other parts of the framework are split across repositories in the Vue.js organization. When building a new framework, it is often easier to manage most of it in a single repository and even keep everything in a single package. The structure can still be well separated, and it will avoid the extra friction of managing extra repositories. As your project grows, it is possible to split it into multiple repositories as you see fit. By that point, the framework should already have an established release model and infrastructure to support this kind of expansion.

To better understand the wider architectural patterns of framework design and maintenance, we are going to explore three important parts that comprise a framework's technical architecture:

- **Modules**: The isolated parts of the code base, usually in a single JavaScript file. These often import or export functions, classes, or other types of structures, which are later utilized as part of a larger system.

- **Packages**: The main source code behind the framework includes interfaces that are exposed to the users and interfaces that are used internally to facilitate certain functionality.

- **Scripts**: Binaries and scripts that are exposed as part of the framework. Some of the scripts are also used for framework development and extra tooling required for specific use cases.

- **Compilers**: The programs that are included as part of the framework that either generate the main JavaScript output from the framework or are used as part of the development process.

Not everything in these categories is required for a successful framework, but as authors, we can pick and choose what is important to us for our project and focus on that. If we look at the existing frameworks of today, we will see similar patterns that work well for JavaScript, frontend, and backend projects – these projects employ all or most of these architectural categories.

In the next section, we will explore two types of technical patterns in framework development – these include architectural and design decisions that affect how the different frameworks function.

Patterns

The focus on framework development requires learning about different types of software patterns, such as architectural, design, and technical patterns. Knowing about these patterns and what kind of decisions went into implementing them in existing frameworks can assist new framework authors at being successful in their projects.

Architectural patterns

We saw examples of existing architectural patterns such as MVC and MVVM in *Chapter 1*. JavaScript frameworks have the freedom to choose any type of architectural model as they see fit for their use cases. For example, the component-based architecture is highly relevant in modern frontend frameworks and is used in many systems, especially those that extend the React library for their feature set. In this pattern, each component is encapsulated with its own state, a view, and a particular behavior, and could even consist of nested components.

In backend frameworks, the middleware architectural pattern is often utilized to manage or mutate incoming requests and outgoing responses. This type of pattern works well due to the nature of the server requests and responses.

A different approach to architectural patterns can be seen within the Electron.js application framework. For the most part, Electron.js is designed on top of process communication between the interface renderer process and the main process for the lower-level operations. This architectural approach does not define a particular name but still directs the architecture toward the separation of concerns. If you look at some of the code bases of the Electron.js applications, you will notice the organization between the two responsibilities of the interface and the backend operations. Frameworks can also use a mix of architectural patterns that combine aspects of object-oriented, functional, and reactive programming to enable the most flexibility. A lot of these concepts can be witnessed within the Nest.js framework, as highlighted in *Chapter 1*.

By browsing the design decisions and the code of various open source frameworks, you can find a variety of implementations of different architectural patterns. As a future framework author, I encourage you to innovate in this space by creating your original architectural patterns or deriving your own approach to the established patterns.

Design patterns

In terms of design patterns, these focus on a much lower level compared to architectural patterns. These patterns address how a framework can solve frequent challenges of organizing code and techniques to enable a cohesive architecture.

In *Chapter 1*, we saw an example of a design pattern via the observer pattern. In addition to the observer technique, frameworks can also utilize a factory pattern, which helps create and manage reusable objects based on some definition. Depending on the implementation and the environment, additional enhancements may include managing the created objects. Another design pattern often seen in all types of JavaScript frameworks is the Publish and Subscribe pattern. This pattern allows both framework internals and components built with the framework abstractions to interact with each other by emitting events and subscribing to those events to create an asynchronous way of communicating between different parts of the system.

Modules and the module design pattern are also universal in all JavaScript software. You will find module APIs defined by the language itself and abstractions around modules refined by different frameworks. This pattern mainly concentrates on enforcing encapsulation, standardizing the code base, and preventing large convoluted pieces of code.

In the next section, we'll look at the technical architecture. It consists of details related to the technical approach to outlining APIs, defining entry points, and using additional tooling to enable framework behaviors.

The technical architecture

The technical architecture and patterns mainly deal with technical challenges. For JavaScript applications, this could mean dealing with rendering a page, responding to a request, interacting with a database, loading an application, and more. In frameworks, the technical challenges extend beyond solving a particular technical problem. Instead, it is about creating a well-designed packaged system that can be beneficial for framework users to build their projects.

To create this system, framework authors need to combine a set of packaged interfaces and a usable set of scripts and also use additional software to improve the JavaScript programming experience.

While learning about the technical architecture, we are going to look at the three categories that enable the essential features of the framework. We are going to explore all the subcategories under these technical topics, as seen in *Figure 3.3*:

Packages	Scripts	Compilers
Core APIs	Binaries and executables	Framework compilers and bundlers
Entry points	File generators	Types
Developer tools		Source maps
Plugin and extension APIs		

Figure 3.3: Subcategories of the technical architecture

We will start with the core packages of the framework that enable substantial functionality.

Packages

Packages consist of the core logic of the framework, with public and private APIs. This directory can include any packages that are necessary for the framework to function. It consists of internal and public interfaces. It can also include a compiler or any building tools that are used as part of framework development or developer use cases. These include the core pieces that make the framework what it is. Depending on the framework, the packages can be functionally independent or rely on each other to function. In general, the packages part of the framework can consist of any related code that should be included as part of the framework, but there are several essential types of packages that today's frameworks include. We are going to look at these in more detail in this section.

Core APIs

Depending on the framework, the "core" packages may consist of a variety of modules for providing public user-exposed interfaces and private interfaces to enable frameworks' features. In *Chapter 2*, we saw examples of public APIs provided by framework packages, such as router, event-management, template modules, and more. The core packages are usually structured similarly in both frontend and backend frameworks.

Frontend frameworks enable **reactivity** features as part of their public APIs. Reactivity enables the components initialized by the framework to update their states as the underlying data changes. When frameworks implement various forms of data binding, they build upon the forms of reactive programming. For instance, Vue.js isolates its main reactivity component in `@vue/reactivity` (`npmjs.com/package/@vue/reactivity`). Reactivity is an important part of a framework feature set, and we are going to explore it further in later chapters.

Dependency injection (**DI**) is at the core of some frameworks and allows modules within the framework and the application to declare their dependencies and use external interfaces. As part of DI, frameworks also have interfaces to declare dependency providers. You will find the use of DI in some frameworks, most famously in Angular. There is a guide and more details on Angular's DI at `angular.io/guide/dependency-injection`. Due to the lack of interfaces and typing in JavaScript, the DI features are not as popular and are enabled through TypeScript or other compilers. You will also find injection features in Nest.js on the backend.

You will find a familiar pattern (as shown in the following code block) for DI in both Angular and Nest.js; the `@Injectable` decorator implementation is defined in one of the core packages of the framework at `github.com/angular/angular/blob/main/packages/core/src/di/injectable.ts`:

```
// service: book.service.ts
import { Injectable } from '@angular/core';
@Injectable({
  providedIn: <root>
})
export class BookService {
  constructor() { }
}
// module: book.module.ts
import { BookService } from './book.service';
@NgModule({
  providers: [BookService],
})
export class BookModule {}
```

Another common core API is the **renderer**, which is responsible for converting the template or the programmatic user interface component into a browser or virtual DOM structure that can be later used on the page or the component output. In Ember.js, rendering is made possible by its internal engine called **Glimmer**. It is included by internal components in Ember and installed from its own source at `github.com/glimmerjs/glimmer.js`. Frameworks may offer flexible rendering options that allow for browser and server-side rendering of components. Some frameworks also allow you to extend and define existing renderers for custom behaviors. Vue.js provides the `createRenderer()` API to do just that – you can learn more about it at `vuejs.org/api/custom-renderer.html`.

Full stack and backend frameworks provide a **server interface** as part of the public APIs. We saw an example of this interface in *Chapter 2*. A good architectural example from that chapter was Nest.js with an option to swap out server implementations. Another good example of a server package is the Next.js server (`github.com/vercel/next.js/tree/canary/packages/next/src/server`). In this case, Next.js uses the server package for a lot of internal functionality. The framework allows some access to the server internals, but it is quite limited. For Svelte apps, the SvelteKit framework takes the approach of exposing adapters. These adapters let the user decide and adapt to their deployment target. For Node.js server users, there is the `@sveltejs/adapter-node` adapter. In addition, the framework provides several official adapters, specified at `kit.svelte.dev/docs/adapters`, that allow you to define adapters for maximum flexibility and compatibility with the deployment environment. It can also be the case that a frontend framework only uses a server as part of its internal components to enable development functionality, but in production use cases, developers would need to provide their own server.

> **Source of the Svelte.js node.js adapter**
>
> The `chapter3` directory provides the source of the Svelte.js node.js adapter after it is extracted by the installation script. The framework uses this to consume code input and produce output in a framework-friendly manner.

Often, all types of frameworks have a **shared package** that could consist of both public and private APIs. Usually, the most general logic, which is reused across many packages, goes into the shared package. Some contenders for the shared package could be code that is related to the runtime environment and the fundamentals. For example, this package can include utility functions, managing and escaping HTML entities, dealing with HTTP status codes, normalizing universal values across the whole framework, and storing constant values usable across the framework.

In the next section, we are going to look at the entry points of frameworks – that is, the glue that connects the core APIs with the start of framework execution.

Entry points

The primary way the framework's users interact with the framework is through an *entry point*. This definition is very similar to an entry point of a simple program, and it is the place where the execution begins when we run a simple computer program. Depending on the framework's abstractions and structure, this kind of entry point could be very different.

In frontend JavaScript frameworks, the concept of the entry point varies. These frameworks can be included via a `<script>` tag and later initialized on the page by calling the entry points. Angular has a root bootstrap module that is called upon to initialize the application, as shown here:

```
app.module.ts
@NgModule({
  imports: [
    // ...
  ],
  declarations: [
    AppComponent,
  ],
  bootstrap: [
    AppComponent
  ]
})
export class AppModule { }
```

In the preceding code, the `AppModule` root module is essential to initialize the application as it loads in the browser. It also defines the space to include top-level imports and service providers to enable external features in the application.

Ember.js uses a similar pattern by defining an instance of an `Application` class to instantiate a new application. This `Application` class extends the `Ember.Application` class and provides an object literal with configuration options. This object is used to configure the various components and features of the application. This entry point class is used to hold other classes of the application as developers further expand the features of their project. To get precise details on this `Application` class, check out the API documentation for it at `api.emberjs.com/ember/release/classes/Application`.

Looking at SvelteKit as a different framework, due to the way it defines its abstractions, it relies on compiler and build tooling to be the initial entry point. The compiler detects the main `page.svelte` file in the root directory and treats that as an entry point to the index page of the application. Unlike Angular and Ember.js, this is a much less verbose entry point.

Another entry point example is the configuration file in Nuxt.js:

```
nuxt.config.js
export default defineNuxtConfig({
   // My Nuxt config
})
```

This file is defined in the root of the project. It allows for framework configuration and extension and accepts a variety of options. All these possible options are available at `nuxt.com/docs/api/configuration/nuxt-config`.

In terms of pure backend frameworks, usually, the entry point is the bootstrap file that starts the server. As the server process boots up, it initializes the server configuration, such as properly binding the process to a particular port. This process is very well illustrated for *AdonisJs* as a state-changing machine at `docs.adonisjs.com/guides/application#boot-lifecycle`. This bootstrap is also similar to *NestJs* as the framework has a server-side `AppModule` that is used to bootstrap the server.

The following source code is for the Nest.js bootstrap script that initializes the application. It has to import `NestFactory` and `AppModule` as part of this process. `await listen` enables listening to incoming requests:

```
import { NestFactory } from '@nestjs/core';
import { AppModule } from './app.module';

async function bootstrap() {
   const application = await NestFactory.create(AppModule);
   await application.listen(process.env.PORT ?
     parseInt(process.env.PORT) : 8080);
```

```
    }
    bootstrap();
```

This entry point file can work with environment variables to bootstrap the application on the desired port. The included `AppModule` file contains the additional modules defined by the user that will be loaded as the script starts. This pattern is prevalent in other frameworks too, such as Express.js and Hapi.js.

Developer tools

You may find sets of developer tools that make it easier to interact with the framework while you're working on applications. These tools are intended to help with profiling, debugging, and other tasks. In frontend projects, these tools are often provided as browser extensions or standalone apps.

Vue.js provides well-integrated tools as part of its workflows through browser and standalone tooling. It allows us to quickly understand the application structure and further debug the output of the application. *Figure 3.4* shows an example of the tools in action:

Figure 3.4: Vue.js browser developer tools

Frontend frameworks provide some good examples of developer tooling, mostly as browser extensions but sometimes as standalone applications to decouple away from browser workflow:

- **Angular DevTools** (`angular.io/guide/devtools`): These tools provide profiling capabilities and debugging features. It renders the component tree, similar to what's shown in *Figure 3.4* in Vue.js.

- **Vue.js DevTools** (`github.com/vuejs/devtools`): As shown in *Figure 3.4*, it provides a detailed component tree with search as well as a route list that helps debug the router configuration. It also enables the timeline view to show a history of interaction with a Vue.js application.

- **Ember Inspector** (`guides.emberjs.com/release/ember-inspector`): This tooling provides an extensive number of features. It provides a way to explore Ember objects, the component tree, routing, data views, and more.

- **React Developer Tools** (`beta.reactjs.org/learn/react-developer-tools`): These tools allow you to inspect components, edit live properties of components, and modify the state.

- **SolidJS Developer Tools** (`github.com/thetarnav/solid-devtools`): These tools visualize and add the ability to interact with the SolidJS reactivity graph. Like other tools, it can inspect the component state and navigate the tree.

An interesting challenge you can consider while supporting developer tools in your frameworks is keeping up with framework updates. Vue.js DevTools approaches this problem by targeting the major version of a framework – it defines a package for each version, such as `app-backend-vue1`, `app-backend-vue2`, `app-backend-vue3`, and so on. Given that most of the time, these are browser extensions, they have a similar architecture for utilizing the `DevTool` extension browser APIs (`developer.chrome.com/docs/extensions/mv3/devtools`).

You may notice the lack of additional developer tools for backend frameworks. In those cases, developers rely on debuggers, such as a text editor or an IDE, as their tools of choice. We looked at Node.js application and framework debugging in *Chapter 2* while debugging a Nest.js application.

In framework development, introducing additional specific developer tools is not a necessary step at first. However, it makes the framework much more pleasant to work with and empowers its users. These days, JavaScript runtime environments generally have great tooling that helps with development no matter what framework you use.

Plugin and extension APIs

In many instances, the packages provided by frameworks also enable extensibility. This kind of extensibility enables plugin and extension development, benefiting the framework in many ways. The provided API empowers other developers to customize the functionality and add new capabilities focused on the specific needs of the applications. It also allows the framework to stay focused on delivering the main feature set and not include every potential feature in the framework's core.

This extensibility may be useful as part of internal development, where framework authors can create adapters and interfaces based on the extension interface. It can also be useful for external use cases, where developers outside of the core development team can create features for specific use cases that plug into the framework. Let's take a look at some examples of plugin interfaces that are provided by the frameworks we looked at in *Chapter 1*:

- **Bootstrap** (`https://getbootstrap.com/docs/5.0/extend/approach/`) has documentation that consists of the guiding principles to create customizable components that work well with its core functionality. For this project, developers defined a set of rules that serve as the guideline.

- **Ember.js** (`cli.emberjs.com/release/writing-addons`) has "add-on" support that can be installed using the CLI with `ember install <addon>`. It has a full section that supports add-on development as part of its documentation. For example, must-have functionality for most applications, such as authentication, is provided via the `Ember Simple Auth` add-on (`ember-simple-auth.com`).

- **Vue.js** (`vuejs.org/guide/reusability/plugins.html`) enables plugin development, which provides new application-level functionality. As part of defining a plugin, developers need to export an `install` function that executes the plugin logic.

- **Angular** (`angular.io/guide/creating-libraries`) calls these extensible interfaces "libraries." Using the `ng` command-line tooling, developers can quickly generate these new libraries. As part of the library's workflow, the new modules get published to *npm* for others to use.

- **Gatsby** (`gatsbyjs.com/docs/plugins`) has several plugins and very good documentation that enables their development. It provides workflows to develop generic plugins for any use, local plugins specific to a single project, and transformer plugins that convert data between types.

- **SolidJS** (`solidjs.com/ecosystem`) defines an ecosystem of add-ons to fit different purposes. Developers can choose from a variety of UI, router, data, and general add-on plugin categories. The most popular category is the user interface additions, which make it easier to approach various web application layouts and widgets.

- **Svelte** (`sveltesociety.dev/tools`) has a set of tools that are developed to improve the bundling, debugging, and editor experience.

- **Hapi.js** (`hapi.dev/tutorials/plugins`) provides a powerful interface to extend the server-side functionality of the framework. The plugins have a fairly simple interface, where an asynchronous function is used to register and perform any additional extended functionality.

- **AdonisJS** (`packages.adonisjs.com`) provides "packages" that extend functionality. To speed up package development, Adonis uses an **MRM** code modification preset, which can be found at `github.com/adonisjs/mrm-preset`. It allows you to quickly scaffold packages for its framework.

From the examples we just discussed, you can hopefully see that as part of writing your framework, it is good to enable this kind of extensibility. It will help the framework grow and benefit all those who are involved in your framework's ecosystem.

In the next section, we'll explore the variety of scripts that help administer today's frameworks.

Scripts

Every framework needs to be able to perform tasks on behalf of its users, and for those building the framework as well, this is where the architecture requires the introduction of various scripts and programs that can perform those day-to-day tasks. These scripts often help developers be more efficient and eliminate redundant tasks. Well-defined powerful scripts can also make the framework very pleasant to use. In this section, we are going to take a look at binaries and executables that are shipped with the framework, file generators, and other popular tooling within frameworks.

Binaries and executables

Binaries and script files help with framework development and sometimes serve as an interface for users of the framework. These scripts can include build steps, automation, and other JavaScript-related tasks. In some cases, these can be helper scripts or can be run consistently as part of the coding process. Often, these are written in JavaScript/TypeScript to ensure cross-platform execution and to keep the workflow consistent with the same language.

Today's frameworks have various executable tasks, and you may find short npm commands or full-fledged script files. These executables can be used for the following purposes:

- **Building and publishing**: This involves releasing new versions of the framework. At the beginning of this chapter, in the *Exploring the core architecture* section, we saw examples of multiple framework versions that are generated based on the same source. This is where a good build script that also updates the change log file and creates source version tags can be handy. This workflow can also involve generating static assets and uploading framework artifacts. A simple example of this can be found in Ember.js (`github.com/emberjs/ember.js/blob/master/bin/build-for-publishing.js`).

- **Full test runner**: An executable to run tests or other test integrations. Frameworks have many types of tests, and it is often important to create a script that can set up the test environment and swiftly execute all or just the needed tests.

- **Development workflow**: A script to quickly get started with the development of the framework. Usually, this includes starting a JavaScript bundler, a file watcher, and sometimes a development server.

- **Managing dependencies**: Here, you can install and rebuild dependencies. Given that a lot of the parts of the framework can be in separate packages or repositories, it becomes much more efficient to automate the dependency management process.

- **Linting and code coverage**: Similar to the test runner, linting and code coverage ensure good code quality standards. These tools analyze the source code and track down abnormal usages of the language. Code coverage tools ensure that tests run through all paths of a framework's code.

JavaScript frameworks use the scripts field of `package.json` (`docs.npmjs.com/cli/using-npm/scripts`) to define a set of common scripts. In the existing frameworks, you may find that the list of scripts defined in that field can be very large – Vue.js, for example, has over 30 script commands defined in the core package file. Angular created a tool called `ng-dev` (`github.com/angular/dev-infra/tree/main/ng-dev`) to manage all development tasks.

AdonisJS (`github.com/adonisjs/core/blob/develop/package.json`) has a fairly short script list that also serves as a good example of possible scripts a framework may require. Here, we can see examples of the publishing workflow, testing, linting, and more:

```
"mrm": "mrm --preset=@adonisjs/mrm-preset",
"pretest": "npm run lint",
"test": "node -r @adonisjs/require-ts/build/
  register bin/test.ts",
"clean": "del-cli build",
"build": "npm run compile",
"commit": "git-cz",
"release": "np --message=\"chore(release): %s\"",
"version": "npm run build",
"prepublishOnly": "npm run build",
"lint": "eslint . --ext=.ts",
"format": "prettier --write .",
```

Some of these scripts are shortcuts, such as the `clean` task, and call out to another tool to perform the action.

As part of building your framework, it is a good idea to identify the common tasks you perform while working on the framework and publishing new releases. Once you do that, create a series of well-defined scripts for those tasks. This will result in a much more organized and pleasant development workflow.

File generators

You will often find that tools are provided to generate code for common components of a project. Usually, these "generate" the necessary skeleton files that can be later modified by the developer to add custom business logic. If you add this kind of scaffolding functionality to your framework, then it will allow developers to reduce the repetition of code written by hand and save time by preventing unexpected errors from being introduced by manually writing parts of the project. Commonly, the generators also create test files and configure the test runner, which is another time saver. These generator commands are often bundled with the framework and provided by a command-line interface.

JavaScript frameworks employ this generator pattern by allowing the application to be scaffolded or allowing developers to scaffold additional components as the project progresses. For example, Angular generates code using schematics (`angular.io/guide/schematics`). It has built-in schematics for its entities but also allows developers to author their own schematics. With this **Schematics API**, you can create custom tasks to perform on the application project, including the ability to fully manipulate the files and directories.

In another instance, Next.js offers an application scaffolding tool called `create-next-app` (`github.com/vercel/next.js/tree/canary/packages/create-next-app`), which allows framework users to quickly get started building an application with Next.js:

```
> npx create-next-app@latest
Need to install the following packages:
  create-next-app@13.1.6
Ok to proceed? (y) y
✔ What is your project named? ... framework-architecture
? Would you like to use TypeScript with this project? > No / Yes
```

The Next.js generator is built using common JavaScript modules and just like many web frameworks, it utilizes Node.js for this type of tooling.

Depending on how you intend your framework to be used, you will have to choose what kind of generator functionality you want to provide. If your framework is part of a large internal project and is often not used to create new application projects, then a schematic-like Angular approach would be more suitable.

In *Chapter 1*, we saw that many of the frameworks, such as Gatsby, are static site generators. This is another use case where a framework can rely on site generators based on some file generation tooling. This pattern can benefit in the same manner as with any generator – abstract away complexity, eliminate repetitive tasks, and reduce maintenance.

Compilers

In general computing, a compiler translates some type of source code into another target source code. JavaScript puts quite a twist on this with a huge number of compiler tools available, with many of these tools approaching the challenges of web development in different ways while adapting to the latest architectural trends. As years go by, more and more frameworks are using some sort of compiler for development purposes, enabling these projects to cherish the benefits of these tools. In this section, we are going to cover some examples that are used in frameworks today.

The development improvements and the workflows that these tools enable vastly benefit framework developers. When you are creating a new JavaScript framework, you will most certainly appreciate utilizing these tools.

Framework compiler and builders

The framework structure is usually assembled with a build tool. The goal of this step is to take all the necessary assets of the framework, perform specific optimizations on them, and output a developer-friendly bundle of code that targets a characteristic runtime environment. A JavaScript framework often uses such a tool as its build system. As part of this section, we'll examine some of the possible compiler and bundler options:

- **tsc**: The compiler behind TypeScript, `tsc` is the binary that can be invoked to build and analyze TypeScript files and turn them into JavaScript files.

- **Webpack**: This is a bundler compiler that can multiplex JavaScript and other web development-related files. Due to its popularity and extensive features, webpack has support for many advanced development features.

- **Turbopack**: A Webpack successor written by its authors in Rust, Go, and TypeScript, Turbopack consists of a bundler and an incremental build system. Similar to Webpack, the Turbo toolchain focuses on bundling your development assets into an optimized bundle. Turbo uses the Rust programming language to achieve faster builds, especially for larger projects.

- **esbuild**: Written in the Go programming language, this tool parallelizes the workload to create JavaScript bundles.

- **Babel**: This is a toolchain for transforming and generating new JavaScript syntax into compatible old syntax while focusing on cross-browser support and support for various JavaScript environments. You can include it in the build pipeline of your framework to make it functional and testable in older browsers.

- **rollup**: A module bundler for creating optimized bundles of JavaScript, it has a large ecosystem of configurations and plugins. It is very suitable for framework use due to its low overhead and output flexibility.

- **Parcel**: This is a bundler tool that focuses on zero or minimal configuration. Parcel ships many built-in optimizations and automatically applies transforms for popular source types into JavaScript. Parcel can be used to produce optimized application bundles consisting of both business logic and framework code.

- **Speedy Web Compiler** (**SWC**): This is based on the Rust programming language, and its focus is to speed up the TypeScript compiling step. It's used by Next.js and the **fresh** framework.

Jest – a testing framework that we mentioned in *Chapter 1* – uses Babel to build the testing framework itself. As the framework user, you can also opt-in to specific JavaScript targets by tweaking the Babel environment configuration via `babel.config.js`:

```
module.exports = {
  presets: ['module:metro-react-native-babel-preset'],
}
```

The preset in the preceding code block allows Jest to run in the React Native environment, which is different from usual JavaScript application runtimes.

The framework compiler can be the core piece of technology as it enables all the main features of the system. Svelte-based applications use the Svelte compiler, which takes a `.svelte` file and outputs JavaScript files:

```
App.svelte
<script>
  let bookChapter=3;
```

```
    console.log(bookChapter);
  </script>
```

Given a basic script, the JavaScript output that's produced includes the required Svelte dependencies and the initialized SvelteComponent:

```
import { SvelteComponent, init, safe_not_equal } from
  "svelte/internal";
let bookChapter = 3;
function instance($$self) {
  console.log(bookChapter);
  return [];
}
class App extends SvelteComponent {
  constructor(options) {
    super();
    init(this, options, instance, null, safe_not_equal,
      s{});
  }
}
```

The preceding generated code is the compiler output and moves the instance of console.log to the instance function. After compilation, the script tag is removed, and the code is wrapped into an instance function, which executes the code as part of the App component. This becomes the entry point to an application that's powered by Svelte. Svelte is capable of parsing CSS styling blocks and HTML-like syntax. More examples of the compiler in action can be found at svelte.dev/examples.

If the framework uses a compiler that converts from statically typed code, then you may need to define and expose the defined types. In the next section, we will explore how that works.

Types

Exporting and building with a type system is another part of framework architecture and has become more and more popular in the modern JavaScript workflow. We have seen many frameworks utilize TypeScript for their architecture. TypeScript provides several options to organize these interfaces and types. It defines the implementation in .ts files and the declaration files in .d.ts files. As part of this architecture, the framework declares its TypeScript types for its internal and external files. The external types are available for the framework consumers as part of the documentation.

The frameworks provide their type declarations as part of their published package. For example, the whole of the Solid.js framework is typed (solidjs.com/guides/typescript), just like many other projects, and includes the types directory when it is installed:

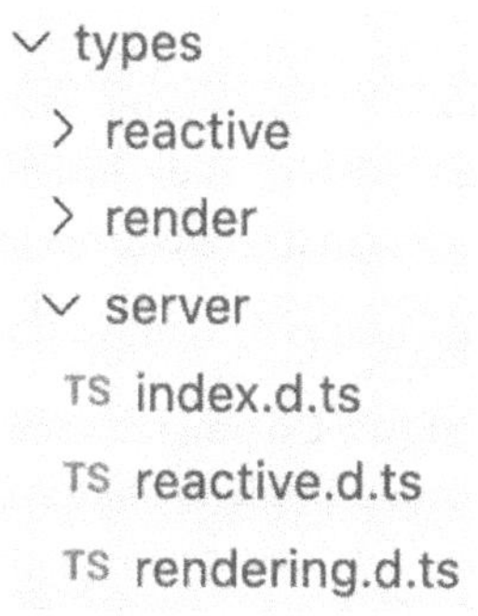

∨ types

  > reactive

  > render

  ∨ server

    TS index.d.ts

    TS reactive.d.ts

    TS rendering.d.ts

Figure 3.5: Provided type declaration in Solid.js

Solid distributes the type definitions as part of the npm package, providing specific definitions for the server, renderer, and reactive interface.

In the next section, we are going to look at source maps, another type of compiled file that gets distributed with frameworks for the benefit of the development workflow.

Source maps

The concept of **source maps** was introduced into JavaScript development more than 10 years ago. The purpose of source maps is to create a map from the built or generated code and turn it into the unbuilt version of the JavaScript file. This kind of map makes it easier to debug the minified or pre-built code. The generated files follow the source map specification (`sourcemaps.info/spec.html`) and usually end in a `.map` file extension. These maps are often created as part of some compilation or build process, hence why we'll learn about them in this section.

Figure 3.6 shows a snippet of a generated source map, though much of it is not humanly-readable:

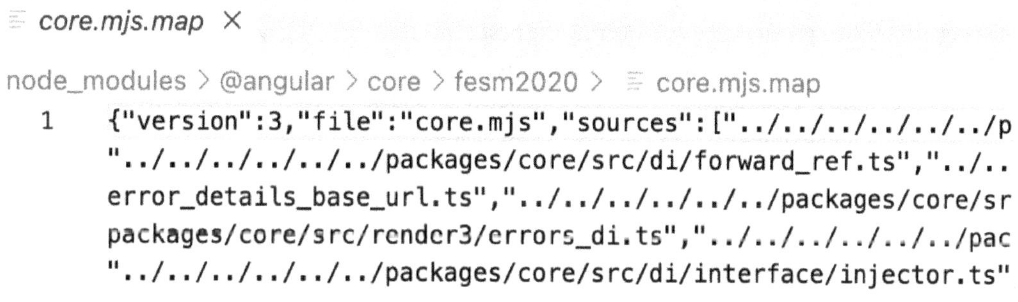

≡ *core.mjs.map* ✕

node_modules > @angular > core > fesm2020 > ≡ core.mjs.map

```
1    {"version":3,"file":"core.mjs","sources":["../../../../../../p
     "../../../../../../packages/core/src/di/forward_ref.ts","..//..
     error_details_base_url.ts","../../../../../../packages/core/sr
     packages/core/src/render3/errors_di.ts","../../../../../../pac
     "../../../../../../packages/core/src/di/interface/injector.ts"
```

Figure 3.6: Generated source map of Angular's core.js file

Here, we can see that the generated map is using the third version of the specification; the rest of the map is meant to be parsed by web-browser tooling that can process most of this file.

As part of the published releases of frontend frameworks, the map files are provided as part of the framework bundle and it is up to developers to decide what to do with these files. The whole project can make the source maps available when the project is deployed into development environments. For example, Vue.js uses *esbuild* to bundle its code in development environments and *rollup* in production builds. As part of the build process, it can pass an option to both of these utilities to generate a source map for the output file. More about this feature in *esbuild* can be found at `esbuild.github.io/api/#sourcemap`. Internally, *esbuild* uses the Go programming language to quickly build JavaScript projects and it has a source map specification implementation in its repository at `github.com/evanw/esbuild/tree/main/internal/sourcemap`. This is just one of the implementations, depending on how you decide to structure your framework. You will be able to find many ways to generate the source maps for your project.

Source maps are utilized in server-side JavaScript as well. Due to the increased use of abstracted ways of writing JavaScript, the source map feature in Node.js can help you trace back to the original code, which could have been written in ClojureScript, TypeScript, or another language.

Usually, it is not that difficult to enable source map support in your framework. However, you need to make sure you configure it properly, making sure that the web browser tools can actually use the map properly and only expose the source map when it is applicable.

Summary

This chapter focused on the technical architectural structure of JavaScript frameworks. We focused on the three important parts of framework architecture: packages, scripts, and compilers. Combining this with the knowledge we gained from *Chapters 1* and *2*, we can start pinpointing the core differences in how various frameworks have their architectures structured. Having an overview of architectural patterns helps us understand how existing frameworks are formulated and makes it easier for us to build new frameworks.

Exploring established projects helps us borrow the best ideas from existing open source frameworks. In addition, learning the internal designs provides insight into how the frameworks fit into complex code bases that utilize a framework. The next chapter will look at development support techniques and patterns that make framework development and usability even better.

4

Ensuring Framework Usability and Quality

Continuing on the theme of framework architecture from the previous chapter, we will begin to look at the more architectural aspects of JavaScript frameworks. While technical architecture plays the core role and provides the guts of the framework, there are additional pieces of system architecture that engineers can add so that the project has a higher grade of usability and quality. As we mainly specialize in JavaScript projects, we will find a variety of tools that help us focus on quality. These tools are often built with JavaScript, but they also integrate with other systems, making it easier for those familiar with the language to cherish the benefits.

Supporting the technical usability of a framework is a series of development quality and usability patterns. These include the additional infrastructure that is used as part of framework development and framework usage. Generally, we look at these components as tools that improve our framework's life cycle. First are the techniques that ensure the framework's usability, for framework authors, contributors, and users. Second is the supporting infrastructure, such as documentation and many types of tests.

We are going to explore these important topics in this chapter, focusing on development support patterns that help us build frameworks. Just like the technical architectures, these skills and tools can be applied while building any type of JavaScript framework. This chapter's topics include exploring the following:

- **Framework documentation** – a set of written or generated materials that provide information on framework features and how to utilize the framework for new projects. To learn from the best, in this section, we will take a look at how other JavaScript frameworks produce public and internal documentation.

- **Variety of framework tests** – used extensively to check the correctness of the framework using different types of tooling, such as unit tests, end-to-end tests, and more. Focusing further on JavaScript projects, this section explores the testing abilities of framework projects.

- **Development tooling** – external tools that help with the development process. This includes additional configurations and tooling that assist with the internal workflow of the framework, such as continuous integration, source control enhancements, and development tweaks. In the upcoming sections, we will see what type of developer tools are used by JavaScript projects such as Vue.js and Nest.js.

- **Generic framework structure** – understanding how we can create our own framework structure based on lessons from other open source framework architectures and patterns that we have seen in the book so far. This will give us a good outline of how authors organize the JavaScript framework code in large projects.

Technical requirements

You can find the resources for this chapter in the book's repository: `https://github.com/PacktPublishing/Building-Your-Own-JavaScript-Framework`. To make it easier to interact with the practical portion of this chapter, use the `chapter4` directory. The interactive script will make it easier for you to run the samples from this chapter. It works as follows from your command prompt or terminal window on your machine:

```
> npm install
...
> npm start
Welcome to Chapter 4!
? Select the demo to run: (Use arrow keys)
> Next.js with Tailwind CSS
  Practical Docus.dev example
```

As with other chapters, the code in the repository is designed to work in operating systems that can run the Node.js runtime, such as Windows, macOS, and most variants of Linux.

JavaScript testing refresher

In this chapter, we will be discussing topics related to JavaScript testing. If you require a refresher or additional resources to give you in-depth information on various types of testing techniques, then check out additional Packt publications at `subscription.packtpub.com/search?query=javascript+testing`.

Development support

Let's dive into a series of techniques and tools that enable developers to build quality frameworks focused on usability and ease of use. Learning and utilizing these types of development support strategies will help us with our framework development, making our projects more usable in internal (work projects) or public contexts (open source/publicly published projects).

Some of these development methodologies and skills are not specific to JavaScript framework development; they are used across many JavaScript and web application development undertakings. However, in the context of framework development, the approach to these tooling and usability patterns is different from a regular application project. For example, a framework might have a particular expanded set of tests that ensures that new features and changes do not break the existing applications that use it. This type of extended testing is a particular case that only applies to framework projects and not application projects.

In addition, a framework may focus more on technical design benchmarking and compatibility testing, supporting a variety of use cases. A lot of this comes from the requirements of the framework consumers and stakeholders.

In the *Core APIs* section of *Chapter 3*, we saw examples of interfaces and features, such as *dependency injection*. These interfaces are designed to empower framework users with their flexibility and feature set. However, these interfaces need to be documented to be accessible to the developers. Otherwise, even if the framework interface is simple enough or powerful but not discoverable or explained, it probably won't be utilized by users. These interfaces also need to be thoroughly tested, in isolation and as part of a greater system. Finally, we need various types of infrastructure to enable this process of testing, maintenance, and documentation. Luckily, there are many existing solutions that make framework development and maintenance much easier through the tools provided by the greater JavaScript community and external software services.

Let us focus on the three categories of usability and quality patterns that will help us build a great framework:

- **Documentation** – a collection of materials targeting different framework stakeholders. These could be generated or written by the framework's developers. This reference can also be internal, showcasing the design decisions and the technical architecture.

- **Framework testing** – the testing infrastructure is crucial to development, feature set, and maintenance as they ensure framework quality. This includes using a variety of tools, such as unit, end-to-end, and integration testing.

- **Development tooling** – the supplementary tools that improve the internal workflow of developers. The tools incorporate technologies that simplify working with the project. They do this by introducing processes such as source control improvements, continuous integration, and others.

Each category of these patterns has several subtypes. In most cases, multiple existing frameworks rely on these types of techniques in their projects. We are going to get into an in-depth look at those patterns in this chapter.

Figure 4.1 presents a detailed outlook at an assortment of documentation kinds, types of tests, and additional tooling that is used by JavaScript frameworks today:

Documentation	Framework testing	Development tooling
Public API	Unit tests	Continuous integration
Examples	Integration tests	Source control
Internal documentation	End-to-end tests	Package configuration
	Benchmarks	

Figure 4.1: Subtypes of tools and patterns utilized in frameworks

We begin by looking at the documentation, something that our framework cannot survive without.

Documentation

Writing up documentation is one of the most crucial things for framework adaptation and usability. The reference materials produced can enable developers to properly utilize the provided tools. Framework developers spent a lot of time writing and tweaking their documentation, focusing on providing the most detailed and simple explanations of the components of the framework function. With a typical web application, you would already have some existing documentation on how to run it and configure parts of it. However, while developing a framework, a lot more documentation is required, consisting of API methods, learning materials, and other solutions to make it easier to utilize the framework to its maximum potential. Today, most frameworks invest in showcasing the framework's API methods, writing up articles that help developers learn the framework from scratch, creating interactive tutorials, and providing detailed examples demonstrating how a framework can approach the challenges of a particular feature set.

A great role model example of documentation comes from the creators of React – with the recently launched new learning platform at `https://react.dev`, which encourages the use of the library across the ecosystem and within frameworks. To help encourage adaption and successful knowledge transfer, their focus was on creating a learning environment alongside the API reference.

As you begin to build out your framework, keep in mind that providing a list of programmatic APIs is not enough. In *Figure 4.2*, we can see excellent examples of valuable reference material:

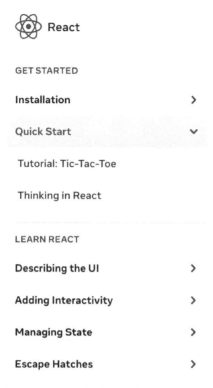

Figure 4.2: Learning React documentation

Even though React is a library, there is still much we can learn about how the documentation is structured for this project. It consists of several vital items. The installation guide comes first and foremost; this can consist of package installation guidelines and the ability to scaffold a new project using a framework. Then, the existing or potential framework user is presented with a tutorial and an explanation of the thinking model behind the tool. Lastly, a series of articles explains the most important topics that a developer needs to know about the tool they are about to use.

Your framework should aspire to a similar form of learning documentation. Even in cases of internal framework development, you or your team should still document and create learning reference material to encourage proper usage of your framework procedures. This kind of approach leads us to the next important part of documentation – the API.

Public API

In *Chapter 3*, as part of exploring framework packages, we have examined the framework entry points. Documentation serves as another type of entry point; developers interact with the framework by utilizing the provided documentation. Primarily, this interaction can be facilitated by the **public API** or **API reference**, created from the framework's code and its interfaces.

Every framework we have seen in this book has an API reference published alongside the framework. This type of API reference can be statically or dynamically generated. In *Figure 4.3*, we see an example of such documentation; the Vue.js docs are generated from the `github.com/vuejs/docs` repository and assembled using a static site generator:

Global API

Application	General
createApp()	version
createSSRApp()	nextTick()
app.mount()	defineComponent()
app.unmount()	defineAsyncComponent()
app.provide()	defineCustomElement()
app.component()	
app.directive()	
app.use()	
app.mixin()	
app.version	
app.config	
app.config.errorHandler	
app.config.warnHandler	
app.config.performance	
app.config.compilerOptions	
app.config.globalProperties	

Figure 4.3: Part of the Vue.js API reference

There are many open source projects that can make it easier for you to generate and maintain documentation files:

- **Docusaurus** (`docusaurus.io`) – a static documentation site generator. Specifically for frameworks, it provides features such as search, versioning according to framework releases, and more.

- **MarkDoc** (`markdoc.dev`) – another open source for custom documentation sites. It is extensible and aims to provide the best documentation authoring and publishing experience.

- **Docus** (`docus.dev`) – a documentation generator that utilizes familiar to us frameworks such as Nuxt.js and Vue.js. Supports Markdown syntax and zero-configuration deploys across many services.

- **TypeDoc** (`typedoc.org`) – a documentation generator for TypeScript source code. Creates static sites based on the comments inside of the TypeScript files. The tool also has the ability to output the parsed source as a JSON file. A similar tool, **JSDoc** (`jsdoc.app`), is also available for pure JavaScript projects.

- **TSDoc** (`tsdoc.org`) – a similar project to TypeDoc, backed by Microsoft. It focuses on standardization efforts around documentation generators by providing an engine for other tools to generate comment-based documentation. Has integrations with several other projects, such as ESLint and Visual Studio Code.

Using one of the tools we just listed might be going overboard if your framework is just starting up, but given that documentation is critically necessary to the framework's usability, you need to make sure to make it easier for yourself to maintain readable and clean documentation. You can also draw inspiration from a larger and more complex framework such as Angular. The project already provides the in-depth API reference (`angular.io/api`), but in addition, the authors also deliver an exhaustive reference to concepts, error types, and diagnostics. All of these can be found in the **guides** section (e.g., `angular.io/guide/file-structure`) of the Angular website.

As you develop your framework, you will likely introduce drastic breaking changes as you go along from version to version. If you are in a context where you already have existing framework consumers, then you should spend the time creating a migration reference document. Good migration guidelines help your stakeholders keep up to date with the changes and use the latest fixes. Some examples of good migration guides include the Electron *Breaking Changes* guides (`electronjs.org/docs/latest/breaking-changes`) and the Express migration overview (`expressjs.com/en/guide/migrating-5.html`).

Practical use of documentation tools

The repository directory for this chapter has an example of one of the documentation tools in action. In this case, the example uses Docus, with the Nuxt.js framework powering the behind-the-scenes infrastructure of the project. You can try this out on your own computer by running the interactive script in the `chapter4` directory or manually navigating to `practical-docus` and running `npm install`, and then `npm run dev`. The documentation site will run on port `3001` and you can live edit and interact with the docs tooling by editing files in the directory.

Basic documentation can improve your framework experience greatly, but there is more you can do to make adaptation more effortless. In the next section, we will focus on the importance of providing examples of framework usage. These meaningful resources can highlight the strong elements of your framework and ease of integration with other systems.

Examples

Providing examples drastically helps reduce one of the most challenging aspects of adapting frameworks – the learning curve. To encourage adoption and reduce friction, framework developers invest time to produce examples of framework usage. These could be included as part of the reference documentation or provided alongside the framework source code. If you are working on an internal framework, investing in examples is still beneficial. If your project will be used by many internal teams or new hires, maintaining a base case of examples can reduce the number of questions and confusion.

As part of my experience contributing to JavaScript testing frameworks, one of the most effective development investments was my focus on creating integration guides and developing examples. This was particularly important for a testing framework project, as the availability of these resources made it easier for developers to add the testing framework into their systems. It also showcased the maturity of the project, showing that it is capable of working with many different systems. It's not just my experience, almost all JavaScript frameworks concentrate on providing instantly-runnable samples. These projects utilize tools such as **StackBlitz** (`stackblitz.com`) and **CodePen** (`codepen.io`) to enable potential developers to get into the framework environment within seconds. For example, navigating to `stackblitz.com/fork/angular` presents you with a ready-to-go Angular framework application.

For more inspiration, Next.js takes the approach to examples thoughtfully; the framework maintains over 50 samples at `github.com/vercel/next.js/tree/canary/examples`. These include showcases of GraphQL support, CMS and blog use cases, integrations with other tooling, and deployment targets. To quickly enable users to run the samples, the `create-next-app` CLI supports the `example` argument to scaffold based on the sample:

```
npx create-next-app --example with-tailwindcss-emotion next-example-
app
```

When developing examples for your framework, keep in mind that you will later need to maintain all the examples you create, just like the documentation reference. If some example code gets out of date and no longer functions as it should, then it will cause more burden for you as the framework maintainer.

Practical example with Next.js

You can check out this example in the `chapter4` directory in the `next-example-app` directory. Follow the `README.md` file for guidelines on setting up the Firebase project. The project requires Firebase project credentials to run properly. To initialize the Next.js app, run `npm install` and then `npm run dev`. You can also use the interactive script directly from the `chapter4` directory.

To open the application, use the localhost URL, which will likely be `http://localhost:3000`. Follow the terminal output for instructions. To edit the files, open the `next-example-app` project directory with Visual Studio Code.

Depending on the nature of your framework, you can use the JavaScript runtime tools to create a playground environment for your framework. In *Figure 4.4*, we see a sample of the Vue.js component playground; this kind of environment takes the idea of examples further:

Figure 4.4: Vue.js single-file component playground

With every "playground" example, you can teach the framework using the most basic features to more advanced use cases.

As you are starting up with your own framework, it is better to include your examples as part of your framework repository. To lessen the maintenance burden, make sure you execute your examples as part of your testing infrastructure (more on that later in *Chapter 10*, related to framework maintenance). If you are working on the framework alone or with a small team, the usage of included examples in your framework can greatly enhance the development process, helping you iterate quickly.

Internal documentation

The internal documentation is all about helping the framework authors to continue developing the framework. Even if you are the only author of the framework, it is still useful to maintain internal documentation, even for your own sake. This way, you can look back on the code and design decisions from the past and make it easier to make updates to your project. Primarily, this type of documentation should not be consumed by the framework users or stakeholders. However, it can still be useful to expose these materials for debugging use cases.

The internal documentation can potentially include detailed interfaces of internal modules. It could potentially describe the design principles of the internal implementation. For instance, the Nuxt.js framework combines both public-facing and internal documentation on its reference pages. The framework's renderer, builders, generator, and other classes are described in the Internals Glossary (`github.com/nuxt/nuxtjs.org/tree/main/content/en/docs/6.internals-glossary`). For instance, Nuxt provides its own module system (`nuxtjs.org/docs/directory-structure/modules`) to extend the framework functionality, and the internals of that feature are supported by the `ModuleContainer` class. This class is part of the framework's internals and should still be documented. It also enables framework plugin development for external developers to understand and extend the framework.

Another example of using this type of documentation for the framework's benefit can be seen in Vue.js. The framework utilizes the TSDoc tooling internally to ensure the specification of its functions, such as the shared utility methods.

The following code taken from the developer tools repository of the framework (`github.com/vuejs/devtools/blob/main/packages/shared-utils/src/util.ts`) is a simpler example of documentation annotations that are available to framework developers while browsing the file, can later be exported to an external document, or can be previewed by an IDE while accessing this helper function:

```
/**
 * Compares two values
 * @param {*} value Mixed type value that will be cast to string
 * @param {string} searchTerm Search string
 * @returns {boolean} Search match
 */
function compare (value, searchTerm) {
    return ('' + value).toLowerCase().indexOf(searchTerm) !== -1
}
```

Contribution guidelines are also part of this type of internal documentation. For both open and closed frameworks, you will potentially have someone who wants to make changes or contribute to the framework, either helping you fix issues or introducing new features. Contribution documentation helps enable this, providing the steps to quickly write and test new framework changes. As part of the contribution instruction, it is often important to list several important pieces of information:

- First, how to modify the framework, build it, and test it. This includes pointers to all the relevant scripts that make the development process more approachable.

- Second, how to successfully write a patch, both in open source and internal environments. This includes following the source control guidelines and commit history rules.

- To make framework contributions easier, this type of documentation should mention the coding rules around public and internal APIs, file formatting, and other potential style guides.

To provide some examples, Ember.js has a page on their contribution guidelines at `guides.emberjs.com/release/contributing`, and other frameworks such as Angular include a `CONTRIBUTING.md` file in their repositories at `github.com/angular/angular/blob/main/CONTRIBUTING.md`.

Framework testing

Just like any software project, frameworks require a number of tests to ensure that the framework is functioning as intended. In the framework context, you will find much in-depth testing focused on correctness, performance, and special framework use cases that need to be handled for all possible usage scenarios. We have seen examples of testing frameworks in the *Testing frameworks* section of *Chapter 1*; those can be used from within our frameworks to simplify the testing workflow. In this section, we are going to look at what techniques JavaScript frameworks use internally to ensure that the final framework product is of high quality.

Unit tests

Just like most software projects, frameworks also include unit tests for their interfaces. They utilize a testing framework similar to the ones we have seen listed in *Chapter 1*. Often, you will see these types of tests called "specs" as well, meaning they are *specification tests*. This means that, given a certain component of a framework, there is a technical specification that it should adhere to. In the framework context, comprehensive tests help with refactoring major components much more quickly. Open source frameworks also benefit from a good unit testing suite when receiving external code contributions. It is much easier to review and be confident in the code change when there is a vast collection of existing tests and new tests being added as part of the change.

Depending on the type of JavaScript framework, the testing environment and testing challenges can be different. In frameworks that target the browser, the unit testing requires mocking out browser and web APIs. For instance, Angular introduces several internal testing interfaces to simplify working with components that are injected into the DOM. Angular's "change detection" and other DOM-related functionality use these testing interfaces to abstract away working with the `document` object instance directly. For example, Angular developers create several test wrappers to make it easier to attach the framework's node tree to the DOM body, as seen in this function:

```
export function withBody<T extends Function>(html: string, blockFn:
T): T {
  return wrapTestFn(() => document.body, html, blockFn);
}
```

The `change_detection_spec.ts` file relies on the `withBody` helper from the test utilities; these utilities rely on executing many frameworks' tests in an environment where a `document` object is present.

In backend frameworks, projects may choose to mock out or create test-only classes based on the existing interfaces. For instance, Nest.js has defined a `NoopHttpAdapter` class (`https://github.com/nestjs/nest/blob/master/packages/core/test/utils/noop-adapter.spec.ts`) that extends `AbstractHttpAdapter`, which we have seen before in the *Backend abstraction* section of *Chapter 2*. The following code shows how the testing adapter is structured to make it easier to use it in framework tests:

```
export class NoopHttpAdapter extends AbstractHttpAdapter {
  constructor(instance: any) {
    super(instance);
  }
  close(): any {}
  initHttpServer(options: any): any {}
  // ...
  }
}
```

This `HttpAdapter` TypeScript class is used within the framework's specification testing to ensure that the main `Application`, `Routing`, and `Middleware` classes function as they should.

While developing your JavaScript framework, ensure to unit test every new component or interface that you add. This process will help you in several ways:

- It will increase the code quality and help you organize your framework such that its components fit better together.

- The framework development process is full of constant refactoring or changes. Your unit testing suite will grow as your framework grows and will increase your confidence in the changes as you code along.

Finally, ensure that your unit testing suite runs efficiently. For example, Vue.js uses the Vitest test runner. Vue has over 2,500 unit tests, which execute in about 20 seconds. The unit tests of your framework should run as fast as possible to provide you with a blazing-fast feedback loop while you are busy developing new framework features.

Integration tests

Integration tests are created with the purpose of testing how multiple interfaces or components of the framework fit together. These tests can catch issues that are not detected by unit/specification testing due to the fact that those types of tests are designed to test the component in isolation. Integration tests simulate the interaction between the components, making sure that the functionality fits well together.

In the context of frameworks, the internal core architecture has to fit together. This means the integration tests would be verifying that behavior. For instance, a good integration test for a full stack framework would be ensuring that a component is rendered when a particular `router` route is called to. This kind of test ensures that all those components are behaving nicely together.

In addition, frameworks usually have to integrate with other systems. This means developers also need to produce integration tests between a framework and that external system. For example, the Gatsby framework has integration tests (`github.com/gatsbyjs/gatsby/tree/master/integration-tests`) for its static-site rendering, command-line interface, and caching infrastructure. These tests verify the framework's features. However, Gatsby also includes integration tests to verify that it works with other technologies. The framework has an integration test to verify compatibility with the JavaScript ESM module standard.

Writing integration tests can be challenging, as you have to verify that all sorts of interface combinations are working faultlessly together. Though it is an essential part of the framework development process, this type of testing can ultimately be much more beneficial than unit testing if you are in a rush to deliver your new framework project.

End-to-end tests

End-to-end tests evaluate how the framework is functioning as a system overall. Usually, these tests simulate almost real user interaction. For a frontend framework, creating these tests usually means configuring one of the end-to-end testing frameworks. For server-side frameworks, the end-to-end tests usually simulate real requests to a server that is powered by the framework. Similar to its set of integration tests, Gatsby also maintains a suite of **end-to-end** (**E2E**) tests (`github.com/gatsbyjs/gatsby/tree/master/e2e-tests`). This helps Gatsby test its full system runtimes and Gatbsy's integration with its theming engine. A simpler example is Vue.js' set of E2E tests for its `transition`, `grid`, and `tree-view` interfaces. These can be found at `github.com/vuejs/core/tree/main/packages/vue/__tests__/e2e`. These tests use Puppeteer to execute commands in a headless Chrome browser, thus simulating real browser and user behavior.

An extensive E2E test suite can support the development of your framework in several ways:

- Catch regressions of the whole system, such as simulating common framework commands and expected functionality.

- Confirm that all components of your framework can work together as you make changes and develop new features.

- Integrate performance trials into E2E tests to be able to detect the slow performance of your framework.

- Ensure that the framework works correctly with external systems. These systems can include different types of web browsers or different backend environments.

Another important form of testing for frameworks that is relevant in many projects today is benchmarking.

Benchmarks

The process of benchmarking runs a set of assessments and trials on a particular scenario of your framework. These benchmark tests can be written by framework authors or an external party. For the purposes of framework building, we are focusing on the former, where a framework includes a series of benchmarks as part of its internal testing. Frameworks can potentially compete on their benchmarking scores for tasks such as rendering a particular component configuration, or, in the case of backend frameworks, the time it takes to process a large number of requests.

For JavaScript frameworks, it is essential to benchmark the performance of the code in the runtime. In the browser runtime, the benchmarks are focused on efficient rendering and processing large inputs. In the full stack Next.js framework, the authors include several benchmarking scripts to test various features (found at `github.com/vercel/next.js/blob/canary/bench`). When you are developing benchmarks of your own, keep in mind that you probably do not need any complex tooling. Instead, you can rely on built-in methods of the runtime – in this case, in Node.js.

Figure 4.5 shows one of the simpler benchmarking scripts `github.com/vercel/next.js/blob/canary/bench/recursive-copy/run.js`):

```
1    async function main() {
2        await createSrcFolder()
3        console.log('test recursive-copy npm module')
4        await run(recursiveCopyNpm)
5        console.log('test recursive-copy custom')
6        await run(recursiveCopyCustom)
7        // ...
8    }
```

```
1    async function run(fn) {
2        async function test() {
3            const start = process.hrtime()
4            await fn(srcDir, destDir)
5            const timer = process.hrtime(start)
6            const ms = (timer[0] * 1e9 + timer[1]) / 1e6
7            return ms
8        }
9
10       const ts = []
11       for (let i = 0; i < 10; i++) {
12           const t = await test()
13           await remove(destDir)
14           ts.push(t)
15       }
16
17       const sum = ts.reduce((a, b) => a + b)
18       const nb = ts.length
19       const avg = sum / nb
20       console.log({ sum, nb, avg })
21   }
```

Figure 4.5: Benchmarking in the Next.js repository

In *Figure 4.5*, the main function provides a recursive copy implementation to the test. Executing the `run` function with two different implementations provides us with comparison results of these two functions.

For backend frameworks, factors such as memory utilization, request throughput, and latency are often benchmarked. For inspiration, Nest.js maintains a set of benchmarking tools to compare the performance of the HTTP servers provided by the framework at `github.com/nestjs/nest/tree/master/benchmarks`. In other types of frameworks, such as application development (Electron) and testing frameworks, the benchmarks are also focused on performance. As we saw in the *Framework testing* section earlier in this chapter, the testing framework itself needs to execute the test suites as efficiently as possible.

You, as a framework developer, should focus on setting up benchmarks for two use cases:

- First, the benchmark for the public interfaces your framework exposes. These would allow you to gauge how long it takes for your framework to complete the tasks.

- Second, you want to look into the micro-benchmarks of the framework internals. These internal benchmarks help optimize particular parts of the framework's core, enabling speed improvements in the internal functions.

As you further develop your project, keep an eye on measurements of your benchmarks, ensuring that you do not regress the speed of your framework.

Development tooling

The framework development and the release process can greatly benefit from the inclusion of additional tools that elevate the quality and usability of the framework project. These workflows could be applied to various aspects of the framework, such as dependency management, testing, editor configuration, formatting, and more. We already saw a similar approach of relying on additional scripts and tooling in the *Binaries and scripts* section of *Chapter 3*. Additional tooling that can improve our framework's development life cycle includes the introduction of continuous integration steps, improving the source control, and the addition of package-level utilities.

Continuous integration

As part of the development cycle, just like many web application projects, frameworks configure **continuous integration** (**CI**) steps to test the code changes and new releases. These CI systems run tests of all types, such as the ones mentioned in the *Framework testing* section of this chapter. Every change committed using a version control system has to pass the existing test suite. This ensures that the changes do not introduce breaking changes or bugs. Besides running the tests, CI runs other types of analysis, such as formatting checks, linting, and more. These ensure consistency, usability, and quality.

If we focus on framework development, there are a few special uses for CI. It ensures that the framework works properly in different JavaScript environments. For a frontend framework, this means executing tests in a variety of browsers on different platforms. Browser support testing goes both ways – new features must work in older browser versions, and new browsers should not break any of the existing framework features. Node.js and Deno frameworks that run on the backend keep track of newer runtime versions, following the `github.com/nodejs/release` release schedule. Running these compatibility checks in CI is the best solution; the CI platforms allow to quickly spin up different versions of environments and parallelize the test execution in those environments.

Besides the focus on testing in various environments, the CI step can run a series of tests of projects that depend on your framework. For example, it can generate and run a sample application or an external script with the new framework changes applied. This way, it can check whether the changes are compatible.

Depending on the framework CI configuration, the integration story may be different. In *Figure 4.6*, we see four successful checks; this is part of the Vue.js CI pipeline:

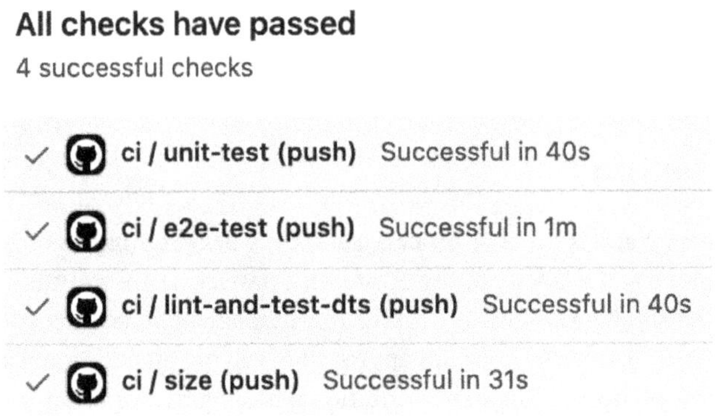

All checks have passed
4 successful checks

✓ 🐙 **ci / unit-test (push)**   Successful in 40s

✓ 🐙 **ci / e2e-test (push)**   Successful in 1m

✓ 🐙 **ci / lint-and-test-dts (push)**   Successful in 40s

✓ 🐙 **ci / size (push)**   Successful in 31s

Figure 4.6: Vue.js reporting its CI status

In the case of Angular, there are over 20 checks in the CI pipeline before a code change can be merged into the repository. The reason for a growing number of integration steps is the execution tasks that are not tests. These could be formatting, spelling, and JavaScript code usability checks.

The CI steps can vary in complexity and type, and they can also contribute to the release process of your framework. No matter what type of framework you are working on, internal or public, it is highly advised to configure a CI step as part of your framework development. This approach will ensure the quality of the code and help you maintain efficiency in your framework development.

Source control

Using source control for frameworks is similar to using it for other application projects. Nevertheless, JavaScript frameworks rely on source control tooling for tagging framework releases and keeping track of feature development branches. The use of source control in this context is a bit more in-depth. For instance, a framework author might need to write patches for an older version of the framework, which means jumping back to an older Git tag to introduce that change. In many cases, large framework refactoring also takes place in a temporary Git branch.

Most JavaScript frameworks also configure supplementary source control scripts that improve the workflow as new features and changes get developed. In *Figure 4.7*, we see the Nest framework using the Git pre-commit hook to execute JavaScript linting scripts:

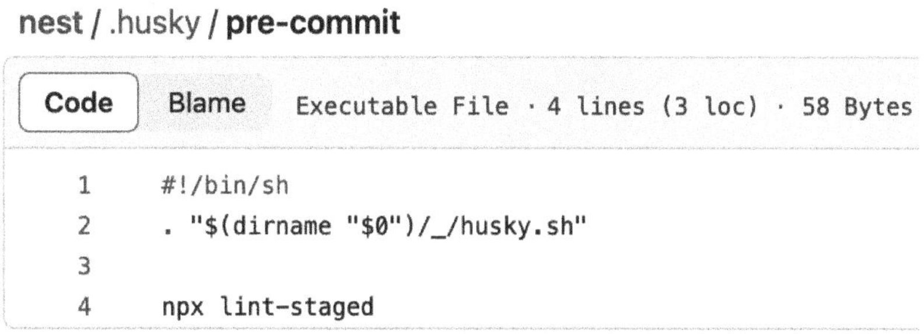

nest / .husky / **pre-commit**

Code	Blame	Executable File · 4 lines (3 loc) · 58 Bytes

```
1    #!/bin/sh
2    . "$(dirname "$0")/_/husky.sh"
3
4    npx lint-staged
```

Figure 4.7: The pre-commit hook configuration

The pre-commit hook, in this case (*Figure 4.7*), enforces the code quality standards before the changes are committed. The step to configure this kind of behavior is simplified using a JavaScript module called **Husky** (typicode.github.io/husky). You will find this pattern in a lot of the frameworks, as this is a handy addition to make the development process much more friendly.

At this point, it is a given that you will use source control for your new framework. However, you can invest in additional tooling by learning from some of the existing frameworks that we've seen in this book to improve your coding workflow.

Package configuration

The package.json file and the additional files in the root of the framework directory define the package configurations of the project. The number of such configurations may vary depending on the types of tools you use in your framework.

The package configuration of Nest.js consists of many tools, such as ESLint, Git, npm, and Gulp, to name a few. Similar to the Nest.js configuration seen in *Figure 4.8*, the package.json file will be the development entry point of your framework:

Figure 4.8: Package configurations of Nest.js

The `package.json` file consists of the information about your framework, its dependencies, and auxiliary configuration used by other tools. For example, the Nest.js package file (`github.com/nestjs/nest/blob/master/package.json`) stores configuration for the `nyc` code coverage tooling, the `mocha` test runner configuration, and commands for the changelog tooling. Besides those configuration entries, the package file has the `scripts` object. This object consists of commands that can be used during framework development. In the Nest.js case, these commands run some of the following operations:

- **Build** – the command that compiles or builds the framework. Nest.js executes the TypeScript compiler as part of this command.

- **Clean** – a quick command that cleans the working directory of the framework project. Usually, this means getting rid of any generated or built files.

- **Test** – commands that run all types of tests that are included in the framework. In the cases of Nest.js and many other frameworks, these types of commands are usually split by the type of tests they are running.

- **Lint** – analyzes the JavaScript code in the project, looking for coding style errors, pitfalls, and potential issues. Nest.js uses ESLint, running in parallel to quickly diagnose the framework files.

- **Publish** – commands that create a new release of the framework. Usually, these include fully building all the parts of the framework, running all the tests, increasing the version number of the project, and finally, publishing the project to npm.

- **Install** – installs the dependencies of the project. In the case of Nest.js, the dependency modules provide the functionality that is required to run a backend service. The developer dependency list has all the infrastructure modules to work on the framework project.

- **Coverage** – runs testing code coverage tooling to determine whether more tests are required to fully cover all the framework logic. For instance, Nest.js uses **Istanbul** (`istanbul.js.org`) for the code coverage report.

This is not an extensive list, but it provides some inspiration for command types you can include in your project. The `scripts` part of the `package.json` file usually follows the reference material at `docs.npmjs.com/cli/using-npm/scripts`, but different JavaScript package managers may treat these commands a bit differently. Your framework should utilize `package.json` to your advantage, creating quickly accessible scripts and configuring the `package.json` file as the entry point for your framework's development workflow.

As we have seen in this section, there are a lot of development tools that empower framework development and are crucial to make the project successful. These development patterns have been improving over many years and are now deeply embedded in many JavaScript projects. In the next section, we shall take a look at the overall picture of a framework structure, which will give us a solid outline for our own framework project.

Learnings from other architectures

In the current and previous chapters, we've seen all the different technical structures, tools, and patterns that frameworks use for their benefit. If we browse through the sources of frameworks gathered in the *Framework showcase* section of *Chapter 1*, we can start to clearly see repeating patterns. Following these practices, we can take advantage of them in our own framework development. By gathering knowledge from the existing designs from different types of JavaScript frameworks, we can conceive a structural system that can serve us well in building our project. We can fuse all those methods and practices into a general framework structure.

The following code shows the generic JavaScript framework structure:

```
<root framework directory>
  | <main framework packages>
    + <core framework interfaces...>
    + <compiler / bundler>
  | <tests>
    + <unit tests>
    + <integration and end-to-end tests>
    + <benchmarks>
  | <static / dynamic typings>
  | <documentation>
  | <examples / samples>
  | <framework scripts>
```

```
| LICENSE
| README documentation
| package.json (package configuration)
| <.continuous integration>
| <.source control add-ons>
| <.editor and formatting configurations>
```

This project structure should help demystify the approach to our framework project structure and empower you, as a developer, to design your own. The framework file and directory structure is the culmination of the last two chapters, combining many of the components that we have seen so far – framework packages, the compiler infrastructure, framework testing, documentation, scripts, and more. Future chapters will be using this structure for our own framework building.

As we look at the core architectures and the examples of what the framework projects consist of, it helps us form what our framework will include and look like. Not everything that we see in the current architecture example will be necessary for our framework to function or succeed. In fact, if you will be building a framework for internal project use only, then you will be choosing a different combination of tools that help you with development.

Summary

This chapter covered the importance of framework documentation, the variety of tests that improve stability, and the array of internal tools that establish efficient framework workflows. The investment in good documentation helps both framework authors and framework users. The lack of well-defined documents can be devastating to the framework's success. Luckily, there are a ton of tools that help simplify the approach to documentation. In the same sense, there are existing tools for testing workflows, covering all aspects of code testing within the framework. Finally, the additional tooling, such as improved source control and editor configurations, makes it much more pleasant to work on the framework and helps authors focus on the framework internals. All of these development support factors play an essential role in framework development and architecture. Learning from other projects and utilizing the patterns that support the development process helps us expand our architectural skills and be more efficient.

At this point, we have learned a lot about existing framework technologies that enable web application and backend service development targeting the JavaScript runtime. In the next chapters, using these learnings and detailed exposure to existing framework projects, we will dive into the aspects of our framework building. This means starting a brand-new project from scratch. Utilizing the patterns, abstractions, and lessons learned from existing projects, we get to experience what it takes to build our own framework.

The next chapter will focus on some of the considerations a framework author needs to take into account before starting off with a new project.

Part 2:
Framework Development

In this part, the book builds on top of the real-world framework examples and switches gears to focus on the programming aspects of framework creation from scratch. The goal is to cover the complete process of planning, architecting, and publishing a new full stack framework. These stages include several important considerations and a lesson in the architecture of various types of components. The emphasis is on the practical approach and guidance of each process's steps that benefit developers of all types.

In this part, we cover the following chapters:

- *Chapter 5, Framework Considerations*
- *Chapter 6, Building a Framework by Example*
- *Chapter 7, Creating a Full Stack Framework*
- *Chapter 8, Architecting Frontend Frameworks*

5
Framework Considerations

In the previous chapters, we primarily focused on learning from other framework projects to prepare for building our full stack JavaScript framework, which will include the ability to create backend infrastructure and frontend interfaces and will have capabilities to test both sides of these features. Even though our goal is a full stack framework for application development, you will be able to use what you learned from this experience and apply it to similar JavaScript projects. Existing projects' architectural patterns and design decisions will help us orient our project and set it up for success. In this chapter, we will study three factors to consider when we plan out our framework that are useful for aspiring software architects and those considering being the responsible individuals behind larger technical project decisions.

Items to cover for the purposes of our framework considerations are set out here:

- **Determining project goals**: Focusing on what you are building and who will be the main consumer and patron of the framework's APIs

- **Identifying framework problem spaces**: Aligning with the problem space of the new framework you are developing

- **Technical design decisions**: Factors such as the technological stack, architecture, and development approach that shape your framework's uniqueness from other projects

In this chapter and generally in this book, we are considering an educational approach to framework building, which means that future chapters will cover the development of specific kinds of JavaScript frameworks, focusing on web application systems. However, you can utilize the gathered knowledge to build a framework that satisfies your particular needs.

The following image will help us focus on the consideration categories and highlights particular subsections that are useful to know as part of the planning and development cycles:

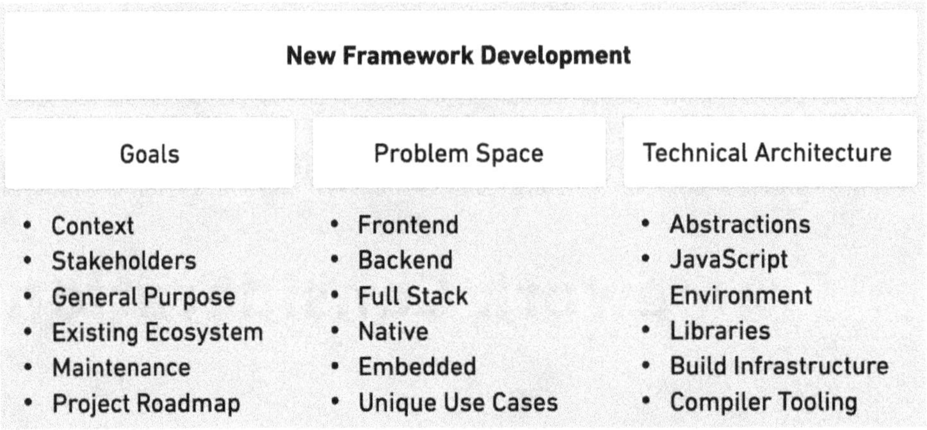

New Framework Development		
Goals	Problem Space	Technical Architecture
• Context	• Frontend	• Abstractions
• Stakeholders	• Backend	• JavaScript
• General Purpose	• Full Stack	Environment
• Existing Ecosystem	• Native	• Libraries
• Maintenance	• Embedded	• Build Infrastructure
• Project Roadmap	• Unique Use Cases	• Compiler Tooling

Figure 5.1: Framework development pillars

Technical requirements

The technical requirements are similar to the preceeding chapters. Use the `chapter5` directory from the book's repository to run `npm install` and then `npm start` to quickly get started with the code and samples that are mentioned in this chapter.

Determining project goals

While contemplating building a new framework, you must identify your project's objectives and stakeholders. These two factors are the main drivers behind your time and investment into building something new. Embarking on a new framework project requires understanding potential motivators and a clear insight into the goals, emphasizing the developer you support and their needs. These reasons can range from internal work use cases to open source hobby projects. Your scenario may very well differ but based on *Figure 5.1*, we can explore the first pillar—*project goals*—as part of this section.

Context

Comprehending the context of a project is fundamental to guiding its development and ensuring it meets the needs of its intended users. The context involves assessing factors such as the project's purpose, target audience, and the environments in which it will be used. Two contexts for framework projects are along the following themes:

- Work-related company-backed initiatives
- Public open source projects, often with an educational or hobby nature

Each of these scenarios comes with its distinctive set of necessities and considerations.

Frameworks developed for particular business needs could have business objectives ranging from small start-up environments to large enterprises. Professional use cases, contextualized in building a framework for business needs, can drastically vary. For instance, similar to the initial development of the React library, a project can support a single platform such as Facebook. However, business needs can also include developing a framework for repeating use cases, such as deploying multiple services with the same core architecture.

Hobby/open source projects can grow from **proofs of concept (PoCs)** or learning materials to widely used tools across the software industry. These are usually not monetarily impactful, but they can serve as a growth tool for your career and practical knowledge. Focus on expanding your skill set in particular software development and JavaScript knowledge areas.

It is a good idea to determine and align the project's goals for your new framework before writing any code. If your project focuses on internal company use cases, such as supporting internal company products, then the goals are more tailored toward enhancing efficiency, improving collaboration, and streamlining the development process. Ultimately, all the planning and building support the main core goal—enabling the organization to ship higher-quality and more reliable projects faster.

Once your framework progresses, the context can potentially alter into something new. For example, if the framework was initially started as an internal project, further down the road, you could open source it and leverage the input from other developers who are not directly associated with it. It can also work the other way—a framework started as a hobby project could be picked up for internal professional use with new investment supported by commercial clients. Both of these scenarios are quite common in the software development industry.

Stakeholders

The target audience and stakeholders of your project are your framework's customers. They are the ones that expect an easy-to-use system with adequate programming interfaces. Catering to their needs and expectations is vital for the success of your framework. These users are looking for a system that is easy to use and provides efficient programming interfaces, allowing them to develop applications and services with ease and agility.

To create a framework that appeals to your target audience, your investment into stakeholder support is essential to ensure that the developer experience is intuitive and user-friendly. Stakeholder support involves designing clear, well-organized resources and providing example code and use cases that solve everyday problems. Often, you will need to offer support to your stakeholders directly or through other means. You will encourage adaptation by making it easier for developers to understand and navigate your framework.

The robust programming interfaces that cater to the diverse needs of your audience also play an important role. By offering practical, adaptable, and efficient tools, you will empower your users to tackle their projects confidently and foster a sense of trust and loyalty toward your framework.

As a framework developer, remember that your audience's needs and expectations should be at the forefront of your design process, so your aim should be to deliver a user-centric experience that stands out in the competitive landscape and becomes an invaluable asset to developers and stakeholders.

In the context of this book, the framework has the reader as the stakeholder, focusing on educational materials. If you are following along and creating your own framework, consider yourself the stakeholder. This makes things much easier; you have the freedom to alter and change the pathway of your framework.

In the next section, we look at several other considerations that could be beneficial to reflect upon before you start coding.

Other considerations

Additional project considerations in framework development are very similar to those of a small or large team developing software projects. Helpful questions to consider as part of determining if your project needs to exist and should be built are along the following themes:

- **The general purpose of the framework**: This is derived from the core objectives; you should have a clear understanding of the primary reasons to deliver your project, identifying specific software-related problems and aiming to solve them.

- **Survey of existing tools**: Before building something new from scratch, evaluating projects already in the ecosystem is a good idea. This consideration will help you make a decision to internally use some of the tools or extend them to fit your needs.

- **Maintenance costs**: Depending on the size of resources to address your framework's problem space, you need to understand time and money investments into the project. Projecting this will help you allocate resources appropriately and ensure the project's long-term sustainability.

- **Innovative and distinctive features**: Identify unique selling points and advantages that your framework will offer compared to existing solutions in the ecosystem. This could include advanced functionality, enhanced performance, or unexplored approaches to solving typical problems.

- **Resourcing**: Assess your development team's skills, expertise, and availability, and identify any gaps that may need to be addressed. This could involve hiring new team members or seeking external support to ensure the successful development of the framework. In some cases, you could be the only resource powering the development of this project. This comes with the advantages of efficient design making but puts you in the driving seat for the whole project.

- **Roadmap**: Develop a comprehensive roadmap outlining project milestones and feature support. By spending time on this, you can provide a clear vision for the development process and help keep the project on track while communicating the framework's direction to its stakeholders.

- **Timeline**: Determine how much time you invest in the framework. This involves setting realistic deadlines for each project stage and considering potential risks and obstacles that may impact the timeline. By establishing a well-defined timeline, you can ensure that the project progresses efficiently and focuses on delivering value to its intended audience.

All the considerations mentioned in this section can contribute to your development process. Spending a little time figuring out the answer to all these considerations could greatly benefit your project. A lot of these considerations will depend on the problem space of your framework. To help handle this, we will cover potential problem spaces in the next section.

Identifying framework problem spaces

Frameworks are designed to support the development of one or many projects, focusing on solving a particular problem space. We define a problem space as a set of challenges or a range of problems that the framework is prepared to address; this is the second pillar from *Figure 5.1*. A problem space can be a specific software application area in which the framework is intended to be used.

As we saw from examples of frameworks in previous chapters of this book, JavaScript has a broad reach in client and server environments. It enables us to build frameworks to fit our needs and technological requirements. For your brand new project, you could potentially be tackling a particular framework category. By focusing on the technical aspects of a unique project, combined with innovative features, you can make your project different from what is already available out there in the wild.

Popular problem spaces

Here are some potential development paths you could take in the JavaScript framework problem space:

- **Frontend frameworks**: Focused on building frontend applications' frameworks, you can find ways to utilize the latest browser technologies to develop unique rendering techniques and improve on state management. Given that this is a popular category of frameworks with many existing solutions, a more straightforward approach is to write your own abstractions on top of these existing solutions—for example, internally using a project such as Vue.js while developing your own self-developed framework interfaces. This way, you can approach your problem space, focusing on its challenges and solving them instead of reinventing the basics.

- **Browser-focused solutions**: Similar to the previous point, you could take the approach of building browser-focused solutions, different from frameworks focused on web application development, and concentrate on utilizing new technologies available to web browsers. For instance, you could leverage WebAssembly (`webassembly.org`) or WebGPU (`w3.org/TR/webgpu`) to develop unique frameworks in the client-side environment.

- **Backend applications**: If you are building a new JavaScript framework for backend development, it is important to focus on reliability, scalability, and security features. You could start by looking over the examples from this book or elsewhere on the internet, then build a framework that addresses common backend challenges, such as working with certain types of databases, enabling different types of authentication, and rapid API development techniques.

- **Testing**: If you are building a new testing framework for JavaScript applications, you could focus on providing a simple and intuitive interface for writing and running tests. You could also enable built-in support for popular testing frameworks and libraries and integrations with continuous integration tools. You could also provide advanced testing features such as visual regression testing, unique parallelization and grouping techniques, and pattern-matching-powered test automation. In *Chapter 6* of this book, we will start developing a simple testing framework, while focusing on the learning process; it is a good contender for a simpler type of framework.

- **Native applications**: By building a new framework for native applications, you could focus on providing components and APIs that make creating responsive and high-performance applications easy. Often, these systems offer built-in support for mobile and desktop features, such as camera access, push notifications, and integration with native operating system features. This type of framework is challenging to develop due to the number of environments you have to support. Though, as we have seen examples in *Chapter 1* and *Chapter 2*, with React Native and Electron, these kinds of projects are not impossible.

- **Embedded solutions**: This type of framework would focus on providing a simple and easy-to-use interface for programming and interacting with hardware devices. To create one of these, you must develop APIs for standard sensors and devices. These include working with external chipsets, motors, GPS, and Bluetooth accessories. The main focus behind this framework would be to create a unique approach to reduce memory and processor usage, as you are targeting embedded instruments. This is a more complex challenge with JavaScript, even though there have been many projects in the past that allowed the runtime to interact with embedded devices.

We explored several examples of more popular JavaScript frameworks in the previous chapters. However, JavaScript's versatility extends beyond those we've already discussed. The language can enable you to build new framework projects that cater to other niches. JavaScript has become ubiquitous in modern web development, allowing developers to build robust and feature-rich applications.

Other framework pathways

There are many other types of projects that JavaScript allows us to build; they all have their own considerations. In no particular order, let's take a look at more framework development pathways:

- **Game development**: As JavaScript is the only target runtime available to the web browser, it ends up being the only solution for building games. Your JavaScript game development framework can provide tools and utilities that make it easy to build 2D or 3D games for web and mobile platforms. These utilities could include built-in support for physics engines, animation, and audio. In such a scenario, the framework could concentrate on providing advanced features, such as multiplayer support or virtual reality integration, as those features become more popular. Some popular JavaScript game development frameworks include Phaser, Pixi.js, PlayCanvas, and Babylon.js. Even though game development frameworks have specific requirements related to rendering and performance, you can still use knowledge from this book to structure this type of framework.

- **Computation**: A new framework for computation and data science could potentially perform scientific computing and data analysis tasks. You could focus on providing a set of APIs for performing mathematical operations and working with data. This framework could have built-in support for popular data visualization libraries and statistical analysis tools. The potential for JavaScript computation frameworks is in the frontend presentation and backend computation layers. A single framework can combine both of those possibilities.

- **Visualization**: Similar to the computation theme, there is room for better JavaScript data visualization frameworks. In visualization projects, you could focus on providing a set of tools and components for creating interactive and dynamic visualizations. You could also integrate with visualization libraries such as D3.js and explore advanced features such as real-time data streaming. As this niche is developing, you can find new ways to render and interact with information from many data sources.

- **Artificial intelligence**: If you decide to develop a JavaScript framework for AI and **machine learning (ML)**, you should prioritize providing a comprehensive set of APIs for building and training neural networks. Consider basing your framework on widely used ML APIs and libraries, such as TensorFlow.js (`tensorflow.org/js`). The features of such a framework could include interoperability with various types of ML formats and configurations.

- **User interface (UI)**: A framework built around constructing UIs could be helpful for a different feature set from regular frontend applications. It could include capabilities for customizable components and responsive UIs. Innovative features could consist of styling and theming components utilizing modern CSS features. The framework could integrate with external libraries such as Tailwind CSS, Material UI, Bootstrap, and so on. Building a UI framework can be advantageous if you work in marketing or design-related environments.

Depending on your framework goal, these are some potential framework pathways. In previous chapters of the book, we have covered the ins and outs of some classes of these frameworks in detail. This is not an exhaustive list of possible solutions within a JavaScript environment, but it showcases the many possibilities. The most popular and competitive category of frameworks is related to building frontend applications.

Now that we have a clearer vision of the problem space, in the next section, we can proceed to considerations with regard to the technical architectures of framework projects.

Technical architecture

In the previous sections of this chapter, we identified our stakeholders—those who will directly benefit from our framework project. We also identified potential problem spaces. Those two factors give us a solid idea of what we want to build. In this section, we explore the third pillar from *Figure 5.1*—*technical architecture*—to give us a focused look at the technical specificities of our planned project.

Abstraction levels and flexibility

The importance of practical abstraction levels and allowing flexibility in code APIs of a JavaScript framework is an important design decision. As you develop your frameworks, these two principles are necessary to ensure the framework's usability, maintainability, and adaptability.

As explored in *Chapter 2*, sensible abstraction levels are essential for providing developers with clean, easy-to-understand interfaces. The encapsulated complexities of the underlying implementation improve productivity and minimize the risk of errors as developers work with a higher-level, more intuitive API that shields them from unnecessary complexity.

The sensible abstraction levels promote modularity and reusability of code, as the framework's features can be more easily connected and adapted to varying circumstances. Providing a level of modularity enables developers to build upon existing modules, fostering a developer-driven ecosystem of extensions that further enhance the framework's capabilities. By striking the right balance between abstraction and flexibility, a JavaScript framework can cater to various projects, from small-scale to complex applications.

Allowing flexibility in code APIs is another critical aspect of a successful JavaScript framework. A flexible API accommodates different coding styles, paradigms, and use cases, enabling developers to tailor their approach to suit their unique requirements. This adaptability is essential in the fast-paced world of web development, where new tools, libraries, and patterns are constantly emerging. By offering a versatile API, a JavaScript framework can remain relevant and valuable in the face of these evolving trends.

One potential pitfall of abstraction is the creation of highly opinionated abstractions that impose strict constraints on how developers can use the framework. While abstractions can streamline specific use cases, they may hinder the framework's overall flexibility and limit its applicability to a broader range of projects. If you would like to build a less opinionated framework, consider providing your stakeholders with expandable options, such as using different templating engines or various ways of managing state within built applications.

Striking the right balance between abstraction and flexibility, and avoiding overly opinionated abstractions, will help you craft a versatile and enduring JavaScript framework.

Environment compatibility

JavaScript runs in diverse environments, including browsers, servers, mobile devices, and other unique hardware, each with unique characteristics, making compatibility a pressing factor in any framework's success. Determining the runtime environment compatibility of your framework is about figuring out which runtimes to support and maintain. Generally, in JavaScript frameworks, it is about the time and technological investment choice of frontend and backend features. This includes browser-specific APIs and compatibility with different backend systems. Besides frontend and backend systems, JavaScript is supported in many other environments.

Framework developers face a significant challenge in ensuring compatibility with multiple JavaScript environments and the specific quirks of those environments. At a high level, this includes different types of browser engines and compatibility with different module systems. This task requires careful consideration and design decisions to ensure the framework works seamlessly across all targeted runtimes.

The first design decision is about configuring compatibility for the appropriate JavaScript environment. Developers must consider the target environment for the framework and ensure that it is compatible with the chosen setting. For instance, if the framework is designed for web applications, the developers must ensure it works seamlessly across multiple browser versions and APIs. Incompatibilities may arise due to variations in browser capabilities, leading to problems such as inconsistent rendering or unresponsive applications.

Another significant consideration when developing a new JavaScript framework is handling environment differences. Writing extra code-compatibility layers is valuable for minor and significant runtime differences. Handling runtime differences includes investing time in backward compatibility for both older browsers and older versions of server-side runtimes. In general, supporting multiple frontend JavaScript environments takes different versions of the same browser environment. For example, many browsers, such as Firefox, have various versions, and each version may have unique capabilities or features. Developers must ensure the framework is designed to handle these variations and provides optimal performance and functionality regardless of the browser version.

For instance, you must handle cross-runtime compatibility to enable server-side rendering or Node.js testing of frontend components. JavaScript server environments may require specific deliberations when building frameworks. They may have different APIs than browsers, and some features, such as the DOM, may not be available. Thus, developers must ensure that the framework is designed to handle such variations and provides optimal performance in server environments. Framework developers include JavaScript polyfills and similar code snippets, providing a fallback mechanism for new and missing features in other environments. These are essential when building a new framework that should work across multiple domains.

Ensuring compatibility with multiple JavaScript environments requires additional thorough testing during framework development and maintenance. Testing is essential in identifying and resolving compatibility issues early in the development cycle. For instance, we can use the automated testing tooling we saw in the previous chapter to test the framework on various browser versions and mobile devices to identify compatibility issues. Including these tests helps to ensure that the framework delivers optimal performance and functionality across all targeted environments. However, testing through all possible runtime use cases and quirks can be challenging, and running a test on all configurations your framework will be used in is impossible. Fortunately, compatibility issues significantly reduced as JavaScript runtimes matured. If you are developing a framework outside the browser, something similar to Electron or React Native, you have further challenges. You must ensure the framework is compatible with the multiple operating systems that you are designing your project for. For example, the operating system runtimes may have different capabilities, affecting the framework's feature set.

Overall, you are able to define the supported JavaScript environments and take control of the types of runtimes you support in your framework, knowing that compatibility with multiple JavaScript environments requires continuous maintenance and updates to your project. This maintenance includes compatibility with new browser versions or server environments that tweak their capabilities or add new features.

Utilizing libraries

It makes sense to commit to specific JavaScript libraries before developing a new framework. The use of existing JavaScript libraries will save you time—time you can use to focus on the framework's features and technical architecture. It is a common pattern for frameworks to rely on libraries to build out the internals. These libraries often indirectly enable the framework feature set behind the scenes, including features such as data management, routing, interacting with the DOM, and abstracting away JavaScript runtime complexity. As the framework covers a more extensive feature set and shapes the development experience, the internal libraries focus on delivering a precise solution to a particular problem.

Choosing the right set of libraries can significantly impact the development process and the shape of your framework. The libraries you utilize in your framework will likely make you an expert user of them. However, balancing the benefits of using libraries with potential downsides, such as compatibility problems, API restrictions, and ongoing maintenance, is necessary.

As we explore other JavaScript frameworks, we can identify libraries they rely on for specific functionality. Depending on the architecture, your framework can build the library right into the framework or use it to extend aspects of your framework. If we look at Angular, we will find that it utilizes **RxJS** (rxjs.dev) for reactive programming, including features such as data observables, iteration, value mapping, and more. The RxJS library can be used directly from within Angular components, such as the following BookService service:

```
import { HttpClient } from '@angular/http';
import { Injectable } from '@angular/core';
import { Observable } from 'rxjs';
import { map, retry } from 'rxjs/operators';

@Injectable()
export class BookService {
  private baseUrl = '...';
  constructor(private http: HttpClient) {}

  getWeather(latitude: number, longitude: number):
    Observable<any> {
    const url = `${this.baseUrl}?latitude=${latitude}
      &longitude=${longitude}&current_weather=true`;
    return this.httpClient.get(url).pipe(retry(3));
  }}
```

The preceding code uses the `Observable` class to return the `getWeather` function first. From within your Angular classes, you can rely upon RxJS to provide many data-operating operators. In addition, the RxJS library provides error-handling operators, as seen in the `retry` call in the preceding code. A detailed explanation of the library's operators can be found at `rxjs.dev/guide/operators`.

Exploring the RxJS example

The `chapter5` directory in the book repository includes an example of using Angular with the RxJS library. You can try this out on your own computer by running the interactive script from the chapter directory or executing `npm install`, and then `npm run dev` from the `angular-rxjs` directory.

The example application will utilize the `BookService` service presented previously to fetch data. The API data comes with additional properties that you can use to extend the existing application. Refer to the `README.md` files for additional information.

In another example of library usage, Vue.js initially used **Vuex** as a library for centralized state management. However, as the framework developed, the approach to managing the state changed. Vue has switched over to recommending and utilizing **Pinia** (`pinia.vuejs.org`) for state management. With an intuitive approach based on the Flux architecture, the library closely related to Vue, it allows developers to use multiple stores to manage states, enables extensibility, and is much more closely aligned with the framework's features. Another example that we have seen in *Chapter 1* is Next.js, which uses the React library for rendering and other features. Next.js focuses on using primitives provided by React to abstract away complexities when using the library directly.

As you introduce libraries into your framework, be smart about choosing them. Often, it is more effortless to abstract away direct access to the libraries for the users of your framework. Otherwise, you must support particular library APIs in your framework, locking you into a specific coding pattern. Historically, Ember.js had to spin up an effort to decouple the framework from its usage of the jQuery library. This type of migration meant providing an update path for projects trying to keep up with the latest versions of the framework.

As your framework develops, you will find great benefits in the ecosystem of JavaScript libraries. The challenge will be keeping up with the developing nature of these projects as the target runtime evolves.

Compiler tooling

In *Chapters 3* and *4*, we examined instances of framework development patterns. These patterns included the use of compilers and other build tools for the purposes of framework development and structure. There's no question that the tools utilized in these patterns make the development, refactoring, and maintenance workflow much more manageable. Therefore, unless there is a specific reason for your framework to avoid the benefits of these tools, it is firmly advisable to lean into the ecosystem.

Build tooling and compilers

While writing the code of your framework project, you want to have a good feedback loop from the code changes you make. This feedback loop can come from running the project's tests or having a sample application that utilizes your framework as you work on new features or bug fixes. This kind of iteration workflow can be configured with built-in JavaScript behaviors, or you could rely on a number of existing build tools and compilers. For the development process, the choice of compiler tooling can significantly impact and affect the efficiency of the development of your framework. Looking back at framework showcases, we have seen examples of using tools such as Rollup.js, webpack, and esbuild for web framework development and packaging.

The choice of these tools will depend on the precise requirements of your framework. While meticulously choosing to use these tools, you need to make sure to evaluate their benefits and drawbacks. In addition, the tools you choose should be suitable for both the development workflow and a good framework publishing workflow. You could decide to separate those two workflows, but then you could end up with too many tools that you need to maintain. For example, we can take some of these tools and consider the following factors:

- The overall JavaScript runtime and feature support, such as features that include extensive JavaScript module format support and advanced features such as tree-shaking and intelligent code bundling
- For frontend systems, evaluate browser and web API support
- The complexity and flexibility of configuration when targeting different workflows and environments, potentially choosing zero-configuration tools versus comprehensive configuration
- Build-time speed for both development and production builds of the framework
- Maturity of the tool compared to other similar solutions
- Developer feature set, such as **Hot Module Replacement** (**HMR**), development server, and instant live reload
- Integration with external tools, such as test frameworks

Some of these factors can differ depending on your framework's problem space—for example, frontend versus backend domains.

Trying out esbuild

The chapter5/esbuild directory in the book's repository includes a sample project that uses esbuild to bundle frontend files. You can refer to the build.js file for the esbuild compiler configuration. When you run this project locally on your machine, the tool will take the assets from the src directory and output the resulting files into the dist directory; these are later loaded into the index.html file in the root of the project. The build steps are executed using the npm run dev command from the project directory.

With enough time investment, you can develop your own compiler or bundler. We have seen prior examples of custom framework compiler development with projects such as Svelte. Creating your own tooling is a larger undertaking, but this is something that could set your framework apart and has immense potential.

JavaScript extensions

TypeScript and similar tools that extend the JavaScript functionality get a special mention in the design decision section. These JavaScript language extensions have been at the core of framework development in recent years. Even if the popularity of using TypeScript in the framework development workflow might decrease over time, it will likely be replaced by other similar tooling that encompasses benefits not available directly with JavaScript. With TypeScript in particular, framework developers get a productivity boost from extra features such as static typing, interfaces, decorators, namespaces, and much more. All these are highly beneficial for framework development.

Suppose you are unsure about introducing an additional TypeScript workflow into your framework or have a specific JavaScript environment that conflicts with TypeScript's tooling. In that case, you could consider a design decision to opt in for the JSDoc annotation version of TypeScript types. A range of supported types for JavaScript files with TypeScript annotations can be found at `typescriptlang.org/docs/handbook/jsdoc-supported-types.html`. If you don't mind the additional transpiling step and entirely opt into TypeScript's ecosystem, then it can help you with many development hurdles, such as the following:

* Reducing the number of code issues identified at runtime

* **Enabling faster refactoring of your code**: The ability to do this is much more critical in frameworks, as frameworks have much more dynamic code bases than routine web application projects

* **Improved class-based programming concepts**: You can use additional building blocks such as interfaces, inheritance features, and more to have a well-designed code base

* **Having a much more descriptive and documented code base**: This proactively benefits you and other teammates working on the project with you

* **It allows you to utilize new syntax features faster**: TypeScript constantly adds new valuable features and is not bound by slow browser adoption of new syntax features

All these benefits are highly useful, and ultimately it is a good design decision to use TypeScript or TypeScript-like solutions to enhance the quality of your coding experience.

Summary

This chapter's core ideas looked at several critical factors and considerations that we need to research and keep in mind as we start our new project. First is identifying the framework's stakeholders and goals, these being the objectives and the audience that will benefit from having these objectives fulfilled. Then, we examined potential problem spaces, focusing on understanding the project types we can consider. Finally, we explored examples of specific JavaScript architectural design decisions that could shape our project.

Considering all this information will help you create a better framework project. Meanwhile, we will also use these framework considerations together throughout this book. We will start applying all these considerations in the next chapter, as we start building a new framework from scratch.

6

Building a Framework by Example

This chapter combines all the insight and architectural knowledge from *Parts 1* and *2* of the book and puts it into practice. Follow along as we develop a simple JavaScript testing framework based on the patterns and best techniques we have seen so far. The practical approach will enable us to learn by example, as it is a great educational approach for software topics of this kind. The framework we build here is a new project developed specifically for this chapter. We can treat this new sample framework project as a "Hello World" exercise for JavaScript framework development. Our aim through this exercise is to train our abilities and apply them later on in real-world projects.

This chapter will cover the following topics for building a framework by example:

- First, we'll structure our initial approach for a new framework project, including determining the goals, stakeholders, and branding for creating something from scratch. This will largely involve putting the learnings of project considerations from *Chapter 5* into practice.

- Next, we'll learn to outline a typical initial architecture design to get our implementation of a testing framework off the ground. This includes outlining how the components fit together and the unique features of the project and its interfaces. In addition, we'll summarize the public interfaces that are expected by developers who will be utilizing our project.

- Finally, we'll walk through the implementation of our testing framework based on the created design, including core feature components, command-line implementation, browser testing integration, and more.

Technical requirements

The implemented framework code is in the book repository at `https://github.com/PacktPublishing/Building-Your-Own-JavaScript-Framework`. The code for our sample framework is included in the `chapter6` directory. To follow the implementation explanation in this chapter, it is recommended to follow along with the framework files in the chapter directory.

To run the code, make sure to have *Node.js v20 or higher* installed (you can download it from `nodejs.org`). To make it easier to manage different versions of Node.js, you can use a Node.js version manager tool such as `nvm` (`github.com/nvm-sh/nvm`). Running `nvm` will automatically install the appropriate version of Node.js from the framework project directory, as the project is configured and tested for a particular version. Make sure to run `npm install` to fetch the project's dependencies. Running `npm test` should output a set of passing tests. This means that your local configuration is all correctly set up. The interactive script will perform all these steps for you if you run `npm start` from the `chapter6` directory.

To test the frontend portion of this sample framework, please use the latest version of Chromium-based browsers or an updated version of Firefox. To debug the framework, refer to the *About debugging* section of *Chapter 2*. You can use the approach as documented in that chapter.

Determining goals and stakeholders and framework branding

The subject of our framework practical exercise is a new JavaScript code testing framework.

Goals

The goal we will try to achieve is to provide robust test tooling for projects of variable complexity. The primary objective of testing frameworks is to provide developers with a reliable, fast, and versatile platform for verifying their code's functionality, performance, and stability. In addition, they aim to minimize the risk of potential errors and create a seamless development experience that ultimately results in a high-quality product.

To accomplish these project goals, our JavaScript framework will focus on executing tests quickly and accurately, supporting various reporter output formats, and fostering a developer-friendly environment. By prioritizing ease of use and integration with other JavaScript tools and application frameworks, we will aim to make the testing process as seamless and efficient as possible. The framework will also enable cross-platform testing across Node.js and real web browser environments, ensuring that code performs consistently and as expected in various contexts.

A side condition of this particular project is to create a sandbox-like environment for you, the reader, by providing an extensible and more straightforward project that you can learn from. You are encouraged to run the code and play the role of a potential framework contributor.

Stakeholders

The stakeholders of a new JavaScript code testing framework could include a variety of software engineers and web developers who have specific software testing needs. First and foremost, JavaScript developers can utilize this framework in their user-facing projects and include it in the development of other JavaScript software. This framework can be a full stack solution for testing code in multiple environments. In the developer community, we are trying to develop a framework that is suitable for frontend, backend, full stack developers, QA and DevOps engineers, and finally, other framework/library authors.

Our stakeholders can benefit from the architectural features of this project. For example, if we are testing the code in a real browser environment, the developed framework could be a good selling point for frontend engineers. The choice of the JavaScript testing feature set that we enable by building this project might also create friction or incompatibility with some engineering workflows, such as running tests in older runtimes and browsers. Thus, mainly, we are targeting a subset of developers that can live on the bleeding edge of new features and easily keep their code base up to date with the latest environments.

If we were developing a JavaScript code testing framework internally, our stakeholders would be the internal developers of a start-up or a bigger company. In this case, we could gather insights about which feature sets were most needed in the internal teams. In addition, particular features could be developed to target specific organizational needs. The iteration and stability of the project would depend on how it was used internally as well.

Both stakeholders and framework developers benefit from consistent project communication and identity. This is where our new framework can benefit a bit from adding some branding. More on that in the next section.

Framework branding

Before diving into architecture and coding, spending some time on the branding and identity of our project can be beneficial for the sake of the project structure. Creating an identity around the framework is a common pattern in internal and public projects, as it clearly defines the project's logical boundaries and provides context around the project in general. Creating an identity could be as simple as picking a temporary name for your new project. It can also be a more involved creative process as we establish a project name. The name will be useful for creating a *namespace* for our project and utilized internally by the code base and by tests/dependent projects as they load this new project. The name can, of course, be changed later with some *find and replace*, so the decision does not have to be final.

We will use a new framework name for our new project and repurpose it for subsequent project advancements in *Chapters 7* and *8*. We define **Componium** as the overarching brand for the frameworks we develop in this book. The name is a play on the word *Component*, combined with a *-nium* ending, which gives this name a technological and software sound. Besides, the name is not fully unique, as a musical instrument from 1821 shares the name with our framework: `mim.be/en/collection-piece/componium`. More importantly, it is not a problem for us because there are no conflicts for this name in the JavaScript ecosystem. Choosing a namespace has to come with some uniqueness, avoiding naming conflicts with existing developer tools.

Continuing the creative process, in *Figure 6.1*, you can see our framework's logo and color scheme, which can be used in documentation and presented in the project's visual features:

Figure 6.1: New framework name and logo

The testing framework will be specifically known as *Componium Test*. The addition of identity assets is beneficial to make the project more identifiable and establish a stronger brand presence in the developer community, fostering trust and recognition.

On the more technical side of things, creating a namespace for your project, as we did here with *componium*, can help on several occasions. Namespacing eliminates potential naming conflicts and collisions with other frameworks and helps us logically group the code base with its components for both internal and external developers. Depending on the type of framework project, creating a namespace can also make the code base approach more modular. We saw an example of this in *Chapter 2*, where Vue.js had the approach to namespace many of its packages in a vue-* namespace. In addition, namespacing can help facilitate framework customizations and extensions, such as other developers contributing plugins and additions.

Now that we know more about creating an identity for our projects and with our branding exercise complete, we can move to the exciting part of architecting our new framework.

Architecting the new framework

Now, we'll get to the architectural part of our practical approach. So far, we know what we are building and who we are building it for. It is now time to determine the shape of our feature set and see how we can enable those features for our users. The most basic use case that we would like to cover is generating assertions for JavaScript code, as presented here:

```
// basic.js
export function helloReader() {
  return "Hello reader!";
}
// tests/basic.js
import ct, { assert } from "componium-test";
import { helloReader } from "../basic.js";
ct({
  basic: function () {
    assert.strictEqual(helloReader(), "Hello reader!",
      "output is correct");
  },
});
```

The preceding code example tests the `helloReader()` function and verifies that the correct string is returned. Starting from these basics, to take things further, we spend time on core functionality and identifying out-of-scope features first. Later, this can help drive technical decisions, as we brainstorm the set of extended features that can be suitable for our project. Our strategy consists of comparing the features offered by existing testing tools, developing unique characteristics, and contemplating which capabilities are out of the initial scope.

To draw some feature comparisons, we can look at some frameworks from *Chapter 1*. *Jest* and *Vitest* come to mind as they are frameworks comparable to what we are building in this chapter. In addition, other projects in the ecosystem are `jasmine`, `ava`, `tap`, `tape`, and `mocha`. Most of these projects provide an advanced testing interface with a specific test life cycle and different ways to create assertions. All these existing open source projects also provide a common set of core functionalities, such as executability to run the framework in different projects, different output formatting options, the ability to stub or spy on interfaces, and more.

In the following subsections, we will examine some of the unique features that can be implemented as part of our initial approach to the *Componium Test* project.

Selecting features

Brainstorming a list of features for a testing framework can be both exhausting and exciting. There are just so many areas to cover in terms of testing tool development, with most of the feature set covering different areas of development. In the architecting example from the beginning of this section, we listed basic test use cases, and now we can expand on them. Here are some of the additional types of functionality that could help our project provide a better feature set to its stakeholders:

- **Capable test runner**: The framework exposes a test runner executable with the ability to execute a single or full suite of tests. It provides its users with a selective execution format to specify which tests to run, a feature that's useful during development and debugging situations. The test runner also allows the users to define test cases using a particular syntax.

- **Cross-platform testing**: The framework enables testing in different JavaScript environments, such as providing some support for Node.js and web API testing.

- **Assertion types**: Users are able to use different types of assertion styles, which provides flexible options for structuring tests. These different assertion styles cater to different preferences and can impact the readability and maintainability of test code. Offering a variety of assertion types allows developers to pick the one that best suits their needs and coding style.

- **Test suite interface**: Provide a rich interface that includes setup functions and life cycle hooks. These features are necessary for organizing and managing tests effectively. A good test suite interface enables necessary actions before and after tests or test suites, enabling a more structured approach to testing.

- **Ability to stub code**: Enable capabilities for spying, stubbing, and mocking of existing code to allow implementation substitution. This is a crucial feature to determine all possible ways the application code can be called. For instance, the Jest framework has the following capabilities built in: `jestjs.io/docs/jest-object#mock-functions`.

- **Code coverage**: Output a report of code coverage that shows a percentage of the code base that is covered by the tests.

- **Integration with build systems**: Expose the right interfaces and document how a particular framework can be executed in **continuous integration** (**CI**) environments, mentioning any particular features that are available to the user as part of this execution mode. Investment into handling build pipeline features can also include integration with build tools (from *Chapter 3*) such as `webpack`, `rollup.js`, and others.

- **Test reporters**: Enable the framework to emit test results reporting in different formats. Some examples of the standard types shipped with frameworks include the following: a specification reporter that shows a hierarchical view of test suite results, a JSON or XML reporter to integrate with external tools, and a dot reporter that outputs . for passing tests or F for failures. Usually, these reporter types can be specified by the environment or flag while running the test framework executable.

- **Plugin interface**: Expose a plugin interface to make it possible to extend the framework feature set. For a testing framework, this could be an option to provide additional reporters or swap out built-in libraries.

Besides this list of features, our framework can also focus on improving the developer experience around JavaScript features such as handling ES module environments or asynchronous behavior in tests. It may also choose to cater to many niche JavaScript use cases, such as being able to work with particular libraries and test media and graphics-related features.

We now have a generous list of conceivable features that will produce a reasonable foundation for our framework implementation. Part of the feature planning and development process also includes figuring out which functionality will be out of scope. This is what we will briefly cover in the following section.

Identifying out-of-scope features

Even though we have identified a rich feature set, it does not cover all potential use cases of a testing framework. To further improve the ergonomics of the project, there are components that would need to be developed in the follow-up versions. In particular, this initial version aims to create a foundation for the framework that will allow us to quickly iterate and add new features. It is better to focus on creating a continuous and intuitive development environment rather than rushing a slew of features immediately. In a real-world use case, as a framework developer, it would be important to gather user feedback and address the most pressing needs of your stakeholders as you iterate on your project.

Here are some features and components that could serve as potential additions to follow-up releases:

- **Test watcher**: Providing a test runner mode that is watching for file changes and rerunning the tests as underlying components change. This could be a welcoming feature for developers as it provides real-time feedback during the development process.

- **Clock and date manipulation**: An additional improvement on top of existing mocking functionality, built-in clock and date mocking in JavaScript projects can make it much easier to freeze or manipulate dates. Providing this interface further improves ergonomics.

- **Snapshot testing**: Involves capturing and comparing the output of complex components. It can save time by simplifying assertion methods and separating the expected output source from the test file structure.

- **Retries**: The ability to retry failed tests could improve the ergonomics in different testing environments. This feature requires a careful approach to make sure that only expected failures are retried and the test runner does not report passing results for broken tests.

There are definitely more components that can be developed for such frameworks, but the feature collection we have identified is enough for us to construct the architectural design outline. The next part of the development process is to pick out the most consequential parts from the feature set and devise an architecture to make this functionality a reality.

Designing the outline

We need to introduce an executable interface to enable the test runner features with cross-platform testing behavior. This interface, implemented as a command-line tool, must handle users' running options within different environments. Finally, to fulfill our main goal of asserting code behavior, the execution modes must return the testing assertions' status.

As the executable triggers the tests, the tests will emit their status and events. The runner needs to respect the life cycles of those tests and subscribe to the events related to the testing workflow. Potential events could include passing or failing state or bubbling up assertion errors if those occur.

Creating a design diagram such as the one seen here makes it easier to see how the components that enable the feature set can interact with one another:

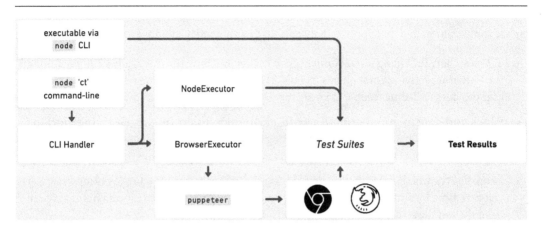

Figure 6.2: Initial design diagram

In *Figure 6.2*, the flow of execution starts from the executable or the command line, defined on the left side of the diagram. In this case, the executable is the ct command, which is short for *Componium Test*. If the developer uses the command line as an entry point, then the **CLI handler** parses through the provided flags and options, determining if the tests will be running in the Node.js environment or the web browser environment. This will enable the cross-platform testing capabilities we evaluated during the feature planning process.

In the Node.js context, the NodeExecutor class can parse the provided test files and run their assertions, logging the real-time status of the execution and outputting the final results of the tests. The chapter6/componium-test/packages/node/executor.js file has the contents of NodeExecutor. The code is programmed to spawn worker environments for each test suite and run assertions in the test cases. Later, a summary of passing and failing tests is collected.

In the browser context, the BrowserExecutor class relies on an external web browser bridge, provided by the puppeteer library. It creates a communication channel between the test runner and a real browser environment, queues the expected test suites, and uses a similar interface to capture test assertion and life cycle output. The browser environment also has distinct running methods, such as the *headless* and *headful* modes, where the browser is either hidden or displayed to the user while the tests run. This feature requires integrating the library and the test runner executable. This executor can be found in chapter6/componium-test/packages/browser/executor.js. This code works in a similar way to the Node.js method, but due to the fact that it runs in the web browser context, it has to send the results of the test suite back to the Node.js test runner running in the terminal process. The environmental difference makes the technical design of these two executors vastly different.

The diagram in *Figure 6.2* also outlines potential support for running tests directly as a script file via the node CLI. In this case, the tests bypass the ct interface and directly output the results as the script runs. It is a handy feature that supports direct testing workflows.

The outline and diagram in *Figure 6.2* do not capture the full details of the test runner workflow, but they are helpful for our development process. As you develop your own frameworks, you will find that creating such artifacts can help you be more efficient and precise in architecting different kinds of software solutions. With the architecture and features outlined, we can continue to the interface design part of the process.

Designing the interfaces and API

Two main interfaces are used as the point of interaction with our framework: the main framework module and the test runner executable. Developers must be acquainted with these interfaces through API documentation, examples, and more. In addition, introducing breaking changes or modifications to these APIs' functionality will affect projects already integrated into the framework.

The first interface we are looking at is the imported `componium-test` framework module. This module exports the main testing object that can accept the testing function through **JavaScript Object Notation (JSON)**. The module also exports additional framework interfaces, such as the assertion and mocking libraries. Here's an example of using this module:

Code snippet 1

```
import framework interfaces
import ct, { assert, fake, replace } from "componium-test";
import Calculator from "../fixtures/calculator.js";
  // fixture to test
let calc;
ct({
  describe: "Calculator Tests",
  beforeEach: () => {
    calc = new Calculator();
  },
  multiply: function () {
    assert.equal(calc.multiply(3, 2), 6, "3 * 2 is 6");
  },
  mockMultiply: function () {
    const myFake = fake.returns(42);
    replace(calc, "multiply", myFake);
    assert.strictEqual(calc.multiply(1, 1), 42,
      "fake interface is working");
  },
  afterEach: () => {
    console.log("called afterEach");
  }
});
```

The preceding code example of using the module in the test file can be found in the `tests/calc-tests.js` path of the framework. In this code, we see the inclusion of basic assertions for a sample `Calculator` script; those are made possible by importing the `assert` function. Based on the expected feature set, we have the test life cycle methods, such as `beforeEach`. Additionally, mocking functionality that replaces the return point of `multiply` is enabled by the `fake` and `replace` functions.

The second interface is the command-line utility provided when the users install the framework. This is a test runner that enables interaction with a large number of test suites:

```
> ct --help
ct [<tests>...]

Options:
        --version   Show version number                       [boolean]
  -b, --browser     Run the test in a web browser             [boolean]
        --keepAlive Keep the web browser alive to debug tests [boolean]
        --help      Show help                                 [boolean]
```

The executable supports a variety of options to enable the expected feature set. In addition, the `--help` flag can also display available commands and their shortcuts. The `ct` executable accepts one or more test suites as arguments, visible through the `ct [<tests>...]` notation. The presence of the `ct` executable is expected by developers after installing the framework from JavaScript's npm package registry.

We now have knowledge of these two developer-level APIs—the test executable and the test interface. We can now proceed to the implementation of features that power the functionality behind those two framework features.

Implementing the new frameworks

The implementation process guides us further on how frameworks in JavaScript can be created. In this implementation process, we'll see the different entry points of a test runner combined with distinguishable features. This section roughly follows the diagram in *Figure 6.2*, where we outlined the framework design. Now that we know our features and have a rough idea of the project's architecture, let's broadly outline the steps of the internal implementation of this project, as follows:

1. Determine how the designed test suite configuration can be implemented within the Node.js environment.

2. Create internal packages to execute the test suites in different environments.

3. Implement the infrastructure for real web browser-mode testing.

4. Collect results from the test runner and output them to the user.

Following *steps 1* to *4*, we begin our implementation by examining how our tests are structured and executed. In the preceding `Calculator Tests` snippet, we can examine the JavaScript object structure, consisting of a `ct` functional call and accepting a test suite object. Some special properties include describing the test suite name and running life cycle methods such as `before` and `after`. The execution life cycle of these methods can be seen in the following diagram:

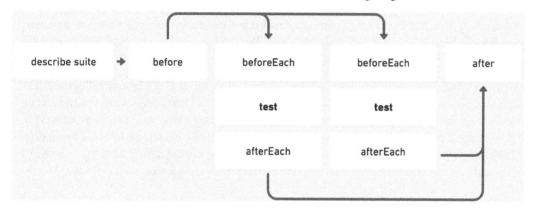

Figure 6.3: Test life cycle

To enable the `ct` function call, the test suites can access the `componium-test` package and our framework's public interfaces, such as the assertion library features. We can see an example of including the framework in the `calc-tests.js` code block in *Code snippet 1*. Besides importing the framework features, the test file imports the code we aim to test.

Importing our framework in the Node.js context will require properly exporting it as part of the package. To do that, we define an `exports` property in our `package.json` file. This allows projects to configure *Componium Test* as a dependency to import the module by name:

Code Snippet 2

```
"exports": {
  ".": "./packages/runner/tester.js",
  "./package.json": "./package.json"
},
```

The preceding configuration pinpoints our `tester.js` file, which handles most of the test-running logic, referenced from our `runner` package.

Once the test object is loaded into the framework, it is directed to the appropriately named `packages/runner/tester.js` file. The `tester.js` file is where most of the testing happens, including the life cycle methods, executed at appropriate times, alongside the main test function. Provided that a test suite can have many tests, the tester proceeds to call through all the provided functions. To measure the speed of the tests, we can rely on the `performance` APIs (`developer.mozilla.org/en-US/docs/Web/API/Performance/now`). These APIs provide high-resolution timestamps to provide us with data on how long the tests take to run.

Testing functionality

Enabling an excellent testing experience through detailed test assertions, our project needs an assertion library that can be used within the newly written tests. The assertion interfaces should include equality comparison, value size comparison, object value evaluation, and others.

To focus on broad framework development, we will import an external test assertion library called **Chai** (`chaijs.com`) to help us with this task. Using this external library can save us time and let us concentrate on developing the framework interfaces and focus on the bigger picture. To avoid locking the project into a particular assertion library, we will abstract away Chai into an internal module. In the initial implementation, we will expose the three interfaces, `should`, `expect`, and `assert`. Subsequently, we can develop our own assertion infrastructure or replace Chai with a different library, which we can potentially develop by ourselves.

As the testing script executes our provided tests, our handy assertion library will throw an exception if any comparison functions are unsuccessful, as follows:

```
AssertionError: 3 * 2 is 6: expected 1.5 to equal 6
    at multiply (file:///Users/componium-test/tests/
      calc-tests.js:20:12)
    at ComponiumTest.test (file:///Users/
      componium-test/packages/runner/tester.js:92:15) {
  showDiff: true,
  actual: 1.5,
  expected: 6,
  operator: 'strictEqual'
}
```

The preceding output shown is a snippet of a failure in one of the tests, where the imported `Calculator` module does not correctly perform the `multiply` operation. The framework needs to handle this type of code behavior. In the case of assertion or other errors, the test runner needs to signal back to the running process of at least one or more failed tests. If at least something flips the `errorOccurred` flag, then it will result in our testing node process exiting with a failure status.

Besides the assertion library, we want to provide mocking and stubbing functionality in the developed framework. Using a similar pattern where we include the Chai.js library and expose it as a test API interface, we will include the **Sinon.js** (`sinonjs.org`) library, which has a rich interface for creating test stubs, spies, and mocks. It is an excellent fit for our project because it supports many JavaScript environments and has been battle-tested in other projects for many years. The presence of this library will make it easier for developers to increase code coverage and write more effective tests.

In the code snippet (*Code snippet 1*) of the example test file, we use Sinon.js' functions to create a fake `mockMultiply` return point. The Sinon.js methods are also abstracted away from direct usage. Instead, they are exposed as part of the `componium-test` package. Users can access the library's mocking, faking, and stubbing mechanisms configured in the framework's `packages/mock/lib.js` mock package.

The inclusion of powerful assertions and mocking libraries, exposed in a simple interface, allows our framework to gain rich functionality from the first version.

Creating a command-line tool

In the previous sections, we talked about making our `tester.js` file the imported entry point to run tests. To match the expected feature set, the framework needs to support running multiple sets of tests from a given directory. This means supporting a number of suites of a given format in bulk. This means that the number of test suites will be structured in the following format:

```
import ct, { assert } from "componium-test";
ct({...});
```

To get this bulk execution feature, we need to create a command-line tool for our framework. To integrate with the Node.js command-line environment, we need to provide a `bin` key in the `package.json` file to export the command-line file, like so:

```
"bin": {
  "ct": "bin/ct.js"
},
```

The binary shortcut that we export is just `ct` to make it really easy to execute the framework commands from other projects. To develop the binary file, we can rely on the `yargs` library (`yargs.js.org`) to handle user input, parsing the provided process arguments from the `process.argv` variable. See the `packages/bin/ct.js` file for details on the executable structure. There are several features to remember with the structure of the executable, focusing on developing a good interface and enabling particular features of the framework, as follows:

- Include the interpretation in the file's first line with `#!/usr/bin/env node`. This signals to the system shell to use the node executable for this file. This is known as the *Shebang* (`wikipedia.org/wiki/Shebang_%28Unix%29`), where an interpreter executable is provided within the script.

- Keep the binary package logic focused on the code related to parsing command-line flags and ensuring a good CLI experience. This includes supporting the `--help` flag, providing shortcuts to flags, and testing for edge cases of how an executable can be used. The flags or options of the CLI should follow the double-hyphenated (`--`) structure or a single one (`-`) for short options.

- The executable should follow the standard rules of CLIs. The CLI process should exit with an appropriate status code and indicate if the process failed. This is especially important in the case of the test runner that we are working with in this chapter.

Depending on the execution mode, Node.js, or the browser, the CLI file (`packages/runner/cli.js`) chooses the right executor class, providing this class with a list of target files. The `NodeExecutor` and `BrowserExecutor` classes have the job of processing all the test files and evaluating if any of the tests in those files fail any assertions. In the Node.js testing environment, we use the `NodeExecutor` class to run the tests. The class spawns new worker threads to execute test suites concurrently. The main purpose of this interface is to run tests for multiple target files and return an overall pass or fail result.

Browser testing workflow

Another feature we are exploring in the developed framework is the ability to execute tests in a real web browser environment. To enable this, the CLI accepts a `-browser` flag that switches the running mode of the test framework. The entry point to this interface for a developer may look something like this: `ct -browser test/some_test.js`. Depending on the approach to the framework executables, we can also introduce a separate `ct-browser` executable to directly execute the test in the browser context without worrying about extra parameters. To implement this functionality, the framework relies on the `puppeteer` library to spawn a new browser instance and establish a communication channel with it.

Figure 6.4 shows the debugging view of the browser test runner workflow:

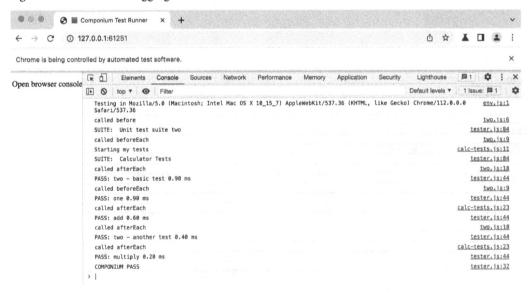

Figure 6.4: Debugging view of the browser test runner workflow

This operation mode is helpful to get access to the tests if they are failing in the browser. This environment allows developers to set breakpoints in their code and debug the tests. The `BrowserExecutor` file in `packages/browser/executor.js` takes care of launching a new instance of a real web browser. The communication channel it establishes can rely on messaging from the console or through more advanced `window.postMessage` function calls.

We can use a simple Node.js server paired with a small templating library called `eta` to create a browser page that executes the tests. We can create a valid HTML document with the templating library that includes the dependencies and test files. To properly load the framework files, the page relies on `importmap` (`developer.mozilla.org/en-US/docs/Web/HTML/Element/script/type/importmap`) to map ES module names to URLs. The code is illustrated in the following snippet:

```
<script type="importmap">
  {
    "imports": {
      "componium-test": "/runner/tester.js",
      "chai": "/chai.js",
      "sinon": "/sinon.js"
    }
  }
</script>
```

The initial implementation of the frontend test runner has a minimal interface, but it can be further developed to show a more interactive dashboard of tests. This can be a potential feature investment, adding additional frontend features if a test runner is used heavily in browser-related testing. The import map structure can be found in the `chapter6/componium-test/packages/browser/views/layout.eta` file. Besides the framework files, this file also consists of parsing a dynamic list of test suites that need to be included in the frontend. This makes it possible to include and run a number of test suites in the frontend testing harness.

Testing the framework

As we develop the testing framework, we must ensure its reliability and correctness. However, using another external framework for testing might not be ideal, and we can't use the framework itself to test its features as it is still under development. We use simple approaches to address this challenge, enabling unit, snapshot, and **end-to-end** (E2E) testing to ensure our project works correctly.

Unit testing of our project can primarily rely on the existing test runner implementation. However, we need to integrate snapshot testing, which includes comparing the test runner's output with snapshotted versions of the desired output. As we refactor certain features, we should verify with the snapshot test to detect unintended changes. Including such tests gives us more confidence as we push new updates and improvements to the project. In particular, unit tests that require more attention are related to the exit status codes of the test runner.

Another important testing addition is the inclusion of E2E tests that cover the whole framework workflow. This includes installing packages, importing the testing package, and executing tests. These tests are made possible by creating a separate testing package that includes our framework as a dependency, as follows:

```
"dependencies": {
  "componium-test": "file:.."
}
```

The sample testing project includes componium-test as a dependency in its package.json file and points to the parent directory to load the files. These E2E tests can be found in the e2e directory of the project. In *Chapter 9*, we will explore additional ways to improve and validate project quality.

Framework packages

Shown in *Figure 6.5* are the initial packages of the project. They are situated in the packages directory, similar to how many other frameworks choose to organize their projects:

```
∨ 📁 packages
   > 📁 assert
   > 📁 bin
   > 📁 browser
   > 📁 mock
   > 📁 node
   > 📁 runner
   > 📁 util
       M↓ README.md
```

Figure 6.5: The framework packages of the project

These packages are logically split by their responsibility. Here is a brief outline of what they do:

- assert: Contains the assertion library-related configuration
- bin: Contains the executable, which is exported by the package.json file
- browser: Files related to test execution in the browser environment
- views: Template files that make it possible to load tests in the frontend context
- mock: Mocking library-related features
- node: Files related to the Node.js execution environment

- `runner`: Global test runner interfaces that are shared between the execution contexts

- `util`: Miscellaneous utility functions that are used across the framework

Besides assembling together all the self-developed packages into a cohesive design, our project also relies on external libraries. Here is a list of some of the dependencies that are utilized in the framework to achieve the desired feature set:

- `chai`: This library allows us to quickly set up a test assertion interface to make the testing experience much nicer. It provides our framework with assertion interfaces useful for **behavior-driven development (BDD)** and **test-driven development (TDD)**. It provides us with multiple assertion styles, and these assertion functions work in Node.js and browser environments. This library has an extensive plugin list that can extend its functionality. The source for this project can be found at `github.com/chaijs/chai`.

- `sinon`: This library allows us to provide an interface for creating test spies, stubs, and mocks. Enabling this feature makes our framework more suitable for testing JavaScript applications, as it enables more comprehensive unit testing coverage.

- `debug`: A small utility that makes it easier to develop the framework by introducing filtered debug logging. By namespacing the debug level after each package of our framework, it makes it easier to understand the framework's internals as it executes tests.

- `eta`: This is a lightweight templating engine that helps us construct the test runner in the browser. It produces an HTML document with the necessary framework files and test suites.

- `glob`: This module enables pattern matching for test directories. It allows us to run a command such as `ct tests` where `tests` is a directory, resulting in finding all test files for a particular directory. Generally, `glob` saves us much time writing filesystem-related code by providing an easy-to-use file pattern matching system.

- `yargs`: This is used as the argument parser for the framework's CLI. It makes it easier for us to create a better command-line experience for the test runner.

- `puppeteer`: This library is included to provide the `BrowserExecutor` class with a real web-browser testing interface. `puppeteer` enables headless browser testing by controlling a Chromium or Firefox instance and running tests within that environment.

These are highlights of the dependencies that we'll use in this project. For some of the framework logic, we have built out our own solutions, organized within packages. At the same time, we are relying on external libraries to address the complexities of specific technical challenges. The current structure allows us to introduce new packages or expand existing ones. The defined abstractions also allow us to swap out external dependencies in favor of other solutions.

Summary

This chapter provided us with a walkthrough on integrating the concepts and architectural principles from earlier sections of the book to build a basic JavaScript testing framework. The practical approach provides us with insights into building out testing framework internals. Even if we do not end up coding a new testing project, it still trains our software muscles to have general knowledge of architecting something from scratch. Our method uses a combination of libraries and packages to enable a hybrid of standard and unique features.

We covered three parts of the framework development workflow: setting up a new project, drafting an initial design, and working through the first version of the goal brainstorming, architectural design, and implementation. As we put these skills into practice, the goal is to make you as a developer more comfortable with architecting, developing, and producing successful projects for others. In the upcoming chapters, we will focus on framework publishing and maintenance. We will also continue these practical exercises as we walk through building more projects.

7

Creating a Full Stack Framework

In *Chapter 6*, we learned a practical example of building a simple JavaScript testing framework. In this chapter, we will continue this practical approach, further diving into the development of frameworks.

The next goal is to develop a full stack framework that will enable developers to build large and small web applications. This chapter will start off by developing the backend parts of such a framework, focusing on the server-side components and integrating essential developer tooling. These backend components, once established, will help us support the frontend elements of the framework we will create in *Chapter 8*. Developing the backend feature set in this chapter will help us do the following:

- Define the technical architecture and goals of our new full stack framework. This is similar to the *Chapter 6* exercise, but now we will switch the context and focus more on the technical challenges of the backend server functionality.

- Learn more about the components that are required to produce functioning full stack tooling. We will study and explore the abstractions we can build and the core parts of the framework that will make it usable in many development scenarios.

- Identify the features that will improve usability, focusing on the features that empower developers and increase efficiency. These comprise tools that help automatically generate the framework scaffold from a template and enhance development productivity.

Technical requirements

The implemented framework code is in the book repository at `https://github.com/PacktPublishing/Building-Your-Own-JavaScript-Framework`. The assets and code are in the `chapter7` directory. As with *Chapter 6*, we will utilize Node.js v20 or higher for this project.

Refer to the `README.md` file of the framework in the chapter directory if you want to run or tweak the framework files locally. `npm scripts` can be handy to use shortcuts during development. As with other projects, to begin working on the framework, you need to install the dependencies with `npm install`.

The framework provides an executable that helps create the scaffold outline of the project, run the newly created application, and more. To locally install the `componium` framework executable from the `chapter7` directory, you can link the executable to use it in different directories. To achieve this, from the checked-out repository directory, use `npm link <path>/chapter7/componium`. This will link a global framework executable to your terminal shell that you can use in a sample project. In addition, if you make any changes to the framework files, the executable will pick up the changes instantly, as the script is directly linked. Check out the detailed `npm link` instructions at `docs.npmjs.com/cli/commands/npm-link` and the following note about Windows compatibility.

A note about Windows compatibility

There are a few things to keep in mind when working with framework executables and commands such as `npm link` in the Windows OS. When running `npm link chapter7\componium` or other commands, you may experience some issues with the executable environment. There are a few common errors that can occur, including the following.

If you get an error related to `enoent ENOENT: no such file or directory AppData\Roaming\npm`, make sure to create that directory in the mentioned path. This is usually an artifact of the npm installer on Windows.

If you get an issue with `UnauthorizedAccessException`, this is a standard security measure of Windows. To fix this, run `Set-ExecutionPolicy RemoteSigned` and allow the execution. For more information, refer to Microsoft's documentation at `learn.microsoft.com/en-us/powershell/module/microsoft.powershell.security/set-executionpolicy`.

You can also refer to the *Common errors* article for npm at `docs.npmjs.com/common-errors`. If you can successfully run `componium --version` using your PowerShell or Command Prompt, then your environment is correctly configured.

Full stack framework goals

Before we get to the software architecture part of this project, we need to get a better insight into what a full stack framework will entail, especially in the JavaScript context and the language's ecosystem. When developing a full stack JavaScript framework, we get to create a blend of abstractions and conventions to help developers produce a combination of frontend and backend components, given the modular nature of the JavaScript language and its ecosystem. From the JavaScript-language point of view, the framework can utilize the latest syntax and functional improvements, such as ES6 modules and modern web APIs. Ecosystem-wise, we will heavily rely on already-established modules to enable certain functionality. Similar to the approach in *Chapter 6*, this strategy allows us to stay focused on the larger system design and achieve much better feature coverage.

The primary objective of a full stack framework is to simplify the application development process, making it faster and more efficient by providing a structured and standardized way to build applications. We have seen examples of the variety of these technical solutions in *Chapter 1*, where frameworks such as Next.js create a much more streamlined and opinionated workflow across the whole stack. The projects, similar to Next.js, encapsulate all aspects of development and cover many use cases, seeking to eliminate the necessity of deciding on different technologies for different layers of development and offer a single unified vision. One of the goals of our sample framework project in this chapter is to focus on similar examples of developing a unified feature set, where components fit naturally together.

The implementation vision of the framework in this chapter should offer a unified API such that the developers utilizing the API can easily get acquainted with its structure and functionality, reducing the learning curve typically associated with new technology adoption. The familiarity and interoperability between different features ensure a seamless experience for developers. The framework API provides abstractions (highlighted in *Chapter 2* in the *About abstractions* section) that make complex operations simple without compromising the flexibility required for advanced use cases. The approach to creating easy-to-use public framework API interfaces is essential to support the varying skill levels of developers and the requirements of robust applications. When you develop or support a full stack framework similar to the practical example in this chapter, you will also have to take a variety of features into consideration. You will find that in your particular scenario, you will invest more in a feature set that is a higher priority for your organization.

Looking at the server-side feature set first, a full stack JavaScript framework must be able to provide a set of fundamental building blocks to handle standard use cases, based on different technological challenges. The features powered by such building blocks could include request routing, database integrations, unified interfaces for events, logging, and performance scaling. In some cases, a framework can also offer features beyond what a developer requires in their projects. Therefore, as part of the development process, we have to strike a balance between using the built-in features and offering a variety of extensibility options. We have seen instances of this in the *Plugin and extension APIs* section of *Chapter 3*, where the server-side frameworks included a way to extend and enable additional functionality. Besides providing a sizeable, flexible feature set with the best developer experience, the applications must be easily deployable and have a way to be monitored once operating in production environments. In this case, pursue a goal of making it easy for developers to deploy their applications, scale those applications up as they get more traffic, and monitor latency and error rates. This involves building in utilities to set up multiple server instances, defining database migrations, and more.

As we define the expected functionality of a full stack framework, we need to define how much the frontend and backend features will interact with each other and where we would like to start the initial development. Considerations for those two sides of a full stack framework are explored in the next sections.

Frontend functionality

On the frontend side of a full stack framework, the emphasis should be on providing seamless integration of frontend components that interact well with backend services. Examples of these types of features could be easily fetching data from the server, rendering frontend components, and having access to the static files. Ember.js, along with EmberData, provides real-world examples of how to effectively collaborate with server-side systems. Another example is Next.js, which provides a tightly integrated solution for both the server and the client. Generally, the opinionated components can define how the framework interacts with the data received from the network requests, and how the data is bound to views and components on the frontend.

The frontend feature set should be capable of handling standard frontend requirements as well. These could include working with popular libraries by offering an easy way to integrate those libraries into projects. An extensive frontend solution should facilitate client-side routing and the ability to manage dependencies and handle form validation, user input sanitization, and JavaScript code bundling. Application state management is another significant component that can help manage the complexity of larger projects.

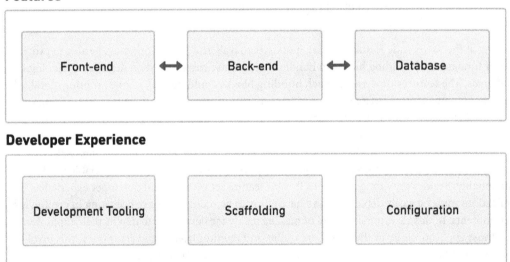

Figure 7.1: The focus of a full stack framework

Ultimately, the process of developing a comprehensive full stack framework calls for the exploration of considerable facets of technology. This initiative also necessitates a broad spectrum of programming skills. You'll be expected to delve into both the complexities of the frontend features, where you will be concerned with user experience, and the backend challenges, focusing on servers, databases, and application logic. The development process enhances your coding abilities and promotes a better understanding of the entire web development ecosystem. In *Figure 7.1*, we break down the two important aspects of our framework development. The **Features** will provide the necessary functionality to build frontend and backend components, interacting with services such as a database, and the **Developer Experience** will make these features reusable and easily accessible for application projects. The result of all this work will be a project usable by a variety of stakeholders, satisfying the respective needs of developers to produce web applications that power new services and web-based software products.

Backend functionality

We start with backend feature development first. This approach will serve the project as a foundation to later add frontend features on top of it to complete the full stack feature set. We will focus on frontend features and architecture in the next chapter. For now, we will imagine that our backend framework can deliver files to the browser client and can be used with any frontend framework. Browser-based applications can make requests to the server-side components to fetch data, static files, or web pages.

Figure 7.2: Our full stack framework logo

We will reuse the same Componium project identity and branding that was established in *Chapter 6*. We now have a new logo, as shown in *Figure 7.2*, to serve as a branding example that identifies our full stack project. The Componium term and branding will cover all parts of the framework will be utilized in the command-line interface as well. Keeping the target developer user base in mind, here is a list of potential starter backend features for this project:

- **API deployment**: The ability to deploy an API service. The API functionality should be flexible enough to allow us to easily configure endpoints and add functional middleware to a selection of endpoints. The API capabilities should also include a modular approach to defining GraphQL resources and schemas.

- **Page rendering features**: The ability to deliver web pages with custom and reusable layouts containing static and dynamic data.

- **Middleware mechanisms**: Features that can add various types of server middleware, such as authentication, to all or some of the routes of the application.

- **Database capabilities**: Database **object-relational mapping (ORM)** capabilities that provide the ability to use different types of database technologies.

- **Application scalability**: Designed to handle high-load situations and offer scalability and performance optimization options. The framework should provide the necessary tools to test, debug, and deploy an application.

- **Development tools**: The framework offers a variety of helpers to ease the development process. The tools can include application bootstrapping and scaffolding of standard components. Live-reloading of the backend server can also save a lot of time by avoiding manually restarting the process on every change.

These functionality goals are now defined in enough detail, covering both the frontend and backend experiences. Utilizing the considerations in this section, we are ready to proceed to the exciting steps of architecting the initial parts of the full stack framework.

Architecture

Now that we have a clearer definition of the backend features, we can start architecting the packages and components of the framework. We will need encapsulation for server and routing interfaces, as well as additional components that allow us to communicate with databases, fetch particular application configurations, and report metrics or logs through logging.

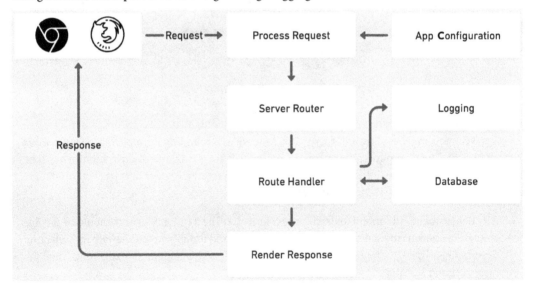

Figure 7.3: The request life cycle

Figure 7.3 represents a simplified view of the request life cycle that our framework will process. The requests can originate in a browser client or, potentially, be direct requests to an API endpoint of the server that our framework needs to handle. You can find a sample application built using our framework in the chapter directory under `tests/sample`.

Figure 7.4: A sample application layout

We now have an overview of how a framework processes requests through its internals. *Figure 7.4* shows an example of an application layout, consisting of the backend API and GraphQL routes. Developers can utilize this structure to build different types of server-side applications. In the following sections, we will take an in-depth look at the entry points, router, configuration, database, and observability features that make this application layout possible. You can find the files in the chapter directory under componium/tests/sample. Some of the files, such as the *.log files, will automatically generate when the application starts and produces some logging; normally, these log files are ignored by the version control system.

Entry point

The server file is the entry point to provide most of the capabilities of our backend framework. It bootstraps a process, loads the appropriate configuration, connects to a database if specified, and establishes the request route handlers. In *Chapter 6*, we had a simpler server (located in the project file at componium-test/blob/main/packages/browser/server.js), developed using the http.createServer(...) Node.js APIs. We can follow the same strategy and build a new server from scratch. However, for a full stack framework, we need a more established solution that has many server interfaces already in place. An excellent middle-ground choice for this project is to use the express module and create abstractions on top of it. *Express* will allow us to define middleware, custom routers, parse requests, craft API responses, and much more. If this sounds familiar, this is because this is a similar architecture to the NestJS framework project previewed in *Chapter 4*.

To expand the server functionality, we will also allow developers to create multiple servers in the same application context, like so:

```
import Componium from "componium";
const app = await Componium.initialize();
const server = await app.createServer({...});
```

With the server interfaces in place, we can now accept requests on a port and configure the servers with options that fit our requirements. Once the server is created, as seen in the preceding code, the main purpose of it is to handle endpoints and perform backend tasks. For the purposes of handling endpoints, we will introduce the *router* interface. It comes into play to accept requests and enable the API features.

Router

In this framework, we want to provide options for file-based and dynamic routing. This involves using a similar feature to file-based routers in Next.js. We saw this concept in the overview of the building blocks in *Chapter 2*. The router implementations can be found in the server package of the framework – for example, the file-based router is in `packages/server/file-router.js`.

The file-based router implementation is capable of recursively loading API endpoints from a directory structure. The file-based feature allows for some manual configuration. By default, the framework can look in a `routes` directory of a project. However, that can be adjusted by passing a file-router directory option to the server configuration. For details on the implementation of a file-based router, refer to the following package – `packages/server/file-router.js`.

As we need to handle different HTTP methods in our file router, we can establish a pattern for the HTTP methods. This way, if you make a POST or DELETE request, you are still able to indicate what type of HTTP method the file should respond to. To enable this functionality, we will configure the framework to handle files in the following manner. Files ending with a `.post.js` will indicate that this route is a POST method route. Files omitting this configuration will default to a GET handler. To handle index or root endpoint routes, we will use an `_index.js` file. This indicates that it is the root handler. For example, the `api/` directory can have the `_index.js` file, which will handle the requests to the root endpoint of the API. Refer to *Figure 7.4* to see the directory structure of the `api/` directory. You can find an example of this in the sample application in the GitHub repository at `componium/tests/sample/routes/api/_index.js`.

Besides file-based routing, developers will sometimes need to define the routes and modular routers directly or programmatically in their code. To support those use cases, our framework offers several server methods to allow flexibility around routes, middlewares, and custom routers, as shown in the following code block:

```
server.addMiddleware("middleware1", () => {
  console.log("Server middleware log");
```

```
});
server.addRoute("/dynamic", function (request, response) {
  return response.send("Dynamic Path");
});
const router = server.createRouter("/router");
router.addMiddleware("routerMiddleware", (request) => {
  console.log("Router middleware:", request.originalUrl);
});
router.addRoute("/sample", function (request, response) {
  return response.send("Router Sample");
});
```

Reading the preceding middleware code, we can see some of the methods that our framework offers to enable dynamic routing. Backed by the primitives from the `express` package, we can create a more abstract system that provides a custom way to describe router definitions. As always, it also allows us to swap out the `express` module in favor of something else later on if required. The `addRoute` methods are present on both the server and custom router objects. These allow us to add custom route handlers instead of using file-based handlers. The `addRoute` function takes an argument of the route path and a function that has request/response arguments, used to process the request. This route-handling flow is visualized in the preceding code, which creates a server middleware, a new route, and an additional router with its own routing entities. Further details on router capabilities can be found at `expressjs.com/en/guide/routing.html`.

In addition to routing methods, we have additional middleware management functions, such as `addMiddleware`. These methods allow us to configure any self-developed or external middleware. In this case, it makes our framework compatible with other express-based middleware. For example, we can use it to add authentication to our endpoints using the `passport` module (`github.com/jaredhanson/passport`); this will enable the framework to support a huge variety of authentication options.

The established architecture of the router and entry points allows us to actually create an application process that can accept requests. In the following subsections, we will define the ability to configure different components of our framework.

Configuration

Developers expect a clean way to configure a framework-based application, with capabilities to run it in local, test, and production environments. To enable this, we will utilize the `config` package (located at `npmjs.com/package/config`). This package will allow us to elegantly organize the configuration options and let the application code base behave differently, based on the environment.

To provide a simple example, if we had only one property to configure, then our configuration file, `default.json`, would look something like this:

```
{
  "database": {
    "connection_uri": "sqlite::memory"
  }
}
```

With the `config` package, we can access the configuration in any of the framework modules and allow it to be used for any developer purposes from within the code base:

```
import config from "config";
console.log("config", config.get
  ("database.connection_uri"));
```

The `config` package automatically loads the required file and allows us to fetch data out of that configuration storage and access all the set properties. In the preceding configuration code, we get to fetch the database connection URL that can be used to bootstrap the database connection from the application.

As with other modules, we can create abstraction on the `config` package. However, we want to offer the core features of the package for developers' use. The `config` package supports several file formats for application configuration, such as YAML (`yaml.org`), JSON, or straight-up JavaScript. Providing a choice of formats is helpful, as it enables us to meet the requirements of the application deployment tools. For example, developer operations might prefer to structure all the production configurations with YAML files.

Configuring aspects of applications within a framework can become remarkably complex, due to the variety of configuration formats and methods available. The JavaScript ecosystem provides dozens of packages that can help you with configuration. Try to evaluate some of them to find the one that fits your needs best. Consider the types of configuration file extensions that a package supports and what kind of validation it can perform. Furthermore, in the following subsection, we will highlight how your framework should offer a set of defaults that work for your users.

Sensible defaults

Sensible defaults are a crucial aspect of a practical application configuration strategy. The baseline configuration inside of a framework makes it easy for applications to work out of the box. Usually, you want to support the most common scenario, such as the development environment. Sensible defaults will allow developers to quickly test-run the framework and see whether they can achieve their project goals by building a prototype. This is especially handy for new developers. Later, the defaults can be overridden to fit the production environment and its requirements. Using the `config` package features will allow us to specify the defaults in the `default.js` file. This file will act as a template for an application's configuration. The configuration file can also be scaffolded in different ways; we will look into this further in the *Developer experience* section.

Reducing the amount of configuration in your framework is something to strive for. It is key to creating a consistent experience and reducing the burdens of building new projects with your framework.

Database

As we add the database functionality to our framework, we need to make sure to support a wide variety of databases. The project should also offer a unified, abstract way to work with database objects. To support these use cases, we can rely on a higher abstract library that enables ORM. For this chapter's example, we will rely on an open source ORM library called **Sequelize** (sequelize. org). This library gives us support for MySQL, PostgreSQL, and MSSQL out of the box. It also enables us to use SQLite in local development and other environments, which makes it really easy to work with database operations and not connect to more complex database services. An ORM library such as Sequelize will add capabilities to interact with a database using JavaScript. It will allow us to treat database entities as objects, simplifying data manipulation and extraction.

In Componium, the database interface can be found in the db package. While some parts of interaction with Sequelize are abstracted, we still rely on the users of our framework to get familiar with the library and its features directly. It is common to see large frameworks rely on separate ORM layers, due to the fact that implementing and supporting several database engines requires a lot of investment in code and time.

When a new application is initialized in code, a framework automatically tries to make a connection to the configured database. To operate with database entities, the application needs to load the model files. In the Componium framework, the models are stored in the models directory and loaded during the initialization as well. This application structure allows framework users to keep data models organized and modular. The models can then be used within your route handlers to interact with the database, whether you need to create, read, update, or delete data. Here's an example of a model we can use in the app:

```
import { DataTypes, Model } from "sequelize";
class Package extends Model {}
Package.init(
  {
    title: DataTypes.STRING,
    address: DataTypes.STRING,
    created. DataTypes.DATE,
  },
  { sequelize: componium.db, modelName: "package" }
);
```

To get access or mutate this model in our route handlers, we can use the `componium` object to get to the database entity object and perform an operation:

```
export default async (req, res) => {
  const packages = await componium.models
    ["package"].findAll();
  componium.logger.info(`Found packages ${JSON.stringify
    (packages)}`);
  res.json(packages);
};
```

The preceding code, located in the application under `sample/api/packages.js`, queries the packages stored in the `Package` model. Finally, it returns all the objects found in a query in the route handler. For the purposes of this chapter, the database implementation is fairly simple, but you can take up the challenge to make the code more accommodating to handle multiple databases and improve model file coupling if desired.

GraphQL support

In addition to the several options to define API endpoints, our framework also supports **GraphQL**, a powerful query language that is useful for backend data retrieval. Some developers might prefer to use GraphQL with our framework, and they should have good experience in integrating this system.

The GraphQL module can be found in the chapter files under `packages/server/graphql.js`. Following a similar design for a file-based router, the `componium` framework has a feature to make it easier to develop modular schemas for GraphQL. These schemas can be defined in separate files and later assembled into full schemas that are supported by every Componium server object. Every GraphQL type definition can be defined in its own `*.gql.js` file. Here's an example of a `packages.gql.js` file:

```
const typeDefs = `#graphql
  scalar Date
  type Package {
    title: String,
    address: String,
    created: Date,
  }
  type Query {
    packages: [Package]
  }
`;
```

First, we define a sample definition for the package type so that we can query the data for that particular model:

```
const resolvers = {
  Query: {
    packages: async () => {
      const packages = await componium.models
        ["package"].findAll();
      componium.logger.info(`Found packages ${JSON.
        stringify(packages)}`);
      return packages;
    },
  },
};
export {typeDefs, resolvers };
```

The preceding code uses the schema definition for a Package type that queries the entries for the objects related to the `Package` model, retrieving all the stored packages in the database. The `.findAll()` call is the built-in interface related to the ORM package that is used under the hood.

Figure 7.5: An Apollo GraphQL sandbox request

The database items can now be retrieved using the query language. *Figure 7.5* shows an interface of the **Apollo Server Sandbox** (`github.com/apollographql/apollo-server`), which allows us to play around with the defined GraphQL schemas. This functionality is available to the users of our framework because Componium utilizes the `@apollo/server` package. GraphQL support can be enabled or disabled by specifying the `gql` option when we create new Componium servers, by calling the `app.createServer({gql: false})` method.

If you are looking to learn more about GraphQL, check out some of the publications related to it at `packtpub.com/search?query=graphql&products=Book`, especially the *Full Stack Web Development with GraphQL and React* book. Many aspects of GraphQL can get really complex quickly, but it is a good example of a feature you can add to your framework to enable a powerful query language based on an established specification.

The next subsection focuses on observability, which will help to make sure the GraphQL feature works properly in applications and we can acquire enough visibility into the internals of the projects using logging.

Observability

The final feature that is worth highlighting has to deal with **observability**. Our framework should offer an interface to allow developers to log essential operations to a file. The framework itself should automatically log any sensitive operations that the developer should be aware of.

This is where the *winston*-based `Logger` class comes into play; you can find the implementation in the framework's `packages/app/logger.js` file. winston (`github.com/winstonjs/winston`) is a logging library that can output the logs in flexible ways and is compatible with various deployment environments. It provides a simple and straightforward interface for logging, allowing developers to easily incorporate logging into their applications with minimal effort. This not only ensures that developers have granular control over what gets logged but also standardizes the way logs are generated across an application.

The logger class is designed to be flexible with respect to the logging environment. It is configured to log to a console in non-production environments, allowing for real-time debugging during the development stages. In production, the logs are written to specified files, providing a permanent record of the application's operation and behavior that can be referred back to when needed. This separation of environments also provides a level of safety, ensuring that sensitive data is not inadvertently exposed during development. These logs can then be used to diagnose and troubleshoot issues, enhancing the observability and reliability of the application. As discussed in the *GraphQL support* section earlier, we will use the logger to keep track of the found packages:

```
componium.logger.info(`Found packages ${JSON.
    sstringify(packages)}`);
```

In development mode, this logger will provide information in the terminal console; this way, this information can effortlessly be seen during the feature development phase. Once the application is deployed and running in the production environment, it will log into the `error.log` and `combined.log` files. The format of the logs in those files will be JSON, such as the following:

```
{"level":"info","message":"Found packages [{\"id\":1,\"title\":\"Paper
Delivery\",\"address\":\"123 Main St.\",\"createdAt\":\"2023-02-01T05:
00:00.000Z\",\"updatedAt\":\"2023-02-01T05:00:00.000Z\"}]","service":"
package-service"}
```

This `logger` class has the interface to log `.info`, `.warn`, and `.error` messages. It will empower developers by providing a way to keep track of the application behavior, debug issues, and store necessary logs that can later be parsed.

Concluding the *Architecture* section, we now have insight into entry points, routing, logging, database, and endpoint support. These are plenty of features for the first iteration of the project. Further, we will connect these features with an improved developer experience.

Developer experience

With several essential backend features built, we can now focus on the developer experience aspects. Earlier referenced in *Figure 7.1*, this is the second important aspect of the framework we will focus on. The goal behind introducing framework-adjacent developer tools is to lessen the friction of adapting the features of our framework. It also helps to streamline the experience and create a unified way to work with the framework's primitives.

To enable most of this experience, we will rely on the `componium` executable that ships with the framework when it is installed. This executable will take care of a lot of mundane tasks, such as initializing applications and scaffolding standard components. It will also eliminate common points of friction by enabling features such as live server reload.

In the next three subsections, we will explore three potential developer experience offerings that you can provide to the users of your framework.

Bootstrapping a new application

The Componium framework's `componium` executable shipped with the library simplifies the process of bootstrapping a new application. Rather than setting up a new application from scratch, the developer can just run the `npx componium init` command. This command will provide options to create a basic project structure with all the necessary configurations and dependencies set up. The initialization process saves a significant amount of time and ensures consistency across different projects built using the framework. The command has the flexibility to accept parameters such as `npx componium init my-app` to create a new application structure in the `my-app` directory. The `init` feature is great for developers, since it automates the repetitive and error-prone process of creating new applications from nothing.

Making this bootstrapping process easy is key for new developers to get started using the framework. Looking back to *Chapter 1*, you will find that almost every framework offers one or several ways of bootstrapping projects, using a CLI tool or a similar mechanism.

Scaffolding

To continue on the theme of reducing error-prone tasks, scaffolding is a highly beneficial feature that further enhances the developer experience. It allows developers to automatically generate boilerplate code for components that are supported by the framework. In the framework exercise of this chapter, we will support creating things such as models, GraphQL schemas, and routes.

```
~/dev/new-componium-app
) componium create
? What would you like to scaffold?
) Model
  Route
  GraphQL Schema
Create a new database model
```

Figure 7.6: An example of the scaffolding experience

Figure 7.6 showcases the developer experience when running the framework executable to scaffold some new components inside of new applications. A highly detailed description can be provided for each option that helps explain the potential options of each operation. The `create` command can also support a direct set of parameters to avoid the `Select` interface for more advanced users of the framework. The scaffolding commands can demonstrate best practices and quickly teach developers about the framework's capabilities.

In addition, the new project initialization scaffolding options can include the framework features a developer wants to use. For example, they can choose the defaults or customize the project and choose their database type, enable GraphQL support, opt in for the file-based router, and so on. To successfully perform this workflow step, the framework has to provide clear instructions related to initializing and setting up the project.

Scaffolding is super common in many projects and is a must-have for new projects. In some cases, the scaffolding features can ship separately from the main project. However, providing the ability to scaffold greatly improves interaction with your framework.

File watching

File watching is another valuable feature provided by the `componium` executable. In this case, it's used for internal usage via the `componium dev` command. This feature watches for changes in the source code files of the application. It later automatically rebuilds and restarts the server whenever a change is detected. This means that developers can see the results of their changes faster and iterate faster. This file-watching infrastructure will come in handy as we further develop the backend features of the framework. It can also be beneficial for the next chapter, where we introduce frontend components that will need to be rebuilt as they are iterated upon.

The combination of these features – easy initialization, detailed code scaffolding, and file watching – creates an excellent developer experience, as it allows developers to focus on the project's requirements. We shall see how these developer features are relevant in the next section, as they relate to an example workflow utilizing our newly developed framework.

The three examples of improvements to the developer experience help to define the specific tasks in which framework authors can invest. These enhancements aim to improve the overall usability of the project. In the next section, the workflow that we will discuss how these scaffolding and file-watching tools can be utilized and accompany the framework project and makes it more pleasant to use.

Developer workflow

In the previous sections of the chapter, we identified the project goals, developed the architecture, and described a few sample features of the framework. In this workflow section, let us summarize how developers can take advantage of the project and understand their workflow.

Here are the parts of the workflow, including some of the steps required on the framework developer side to make the project available for consumption.

Framework distribution

A newly created JavaScript framework is published publicly or privately on the npm package registry. When a framework author wants to keep the project fully private, it can be consumed from a private Git repository within the internal infrastructure. This whole framework package includes a README file and framework documentation that helps external developers start building a new application.

New project requirements

To use an example of framework usage, a developer wants to build a sample application that keeps track of packages. The system needs to be able to list inbound and outbound packages, change the properties of those packages, and add and remove new ones. Ultimately, the developer would like to enable these features and deploy them to a remote environment, where the application interacts with a production/staging database environment.

Starting a project

To start building the application, the developer installs the framework. They can use a command such as `npm install -g componium`. The command globally installs the framework. This is similar to the `npm link` command that we used before to get access to the executable. However, in this case, the framework is downloaded from the npm package manager database. In your case, you can use the npm-linked version of the executable if it makes it easier.

With the framework installed, the developer can now run `componium init` to create a new application structure. The terminal will show the following:

```
> cd new-app
> componium init
Creating /Users/user/dev/new-app/app.js...
...
Installing dependencies...
New Componium application initialized.
```

Using the framework

The initialization script automatically installs the necessary parts of the project, including a `package.json` file. The code scaffolds the necessary files into the project directory based on those choices. The starting point now leaves the developer with a brand-new project directory, where they can begin development. The expectation of this workflow ensures minimal friction during installation and presents the developer with an example of running application code. The next step for the developer is to use either the `componium dev` or `componium create` command. The first command, `dev`, will start the development server in the project directory, and the second command, `create`, can scaffold new components within the project. It is up to the developer to decide whether they want to use the scaffolding helpers or write code from scratch using the provided framework documentation. These two commands will come in handy in the next parts of the workflow, where a developer can start adding new API models and server endpoints.

Creating an API and a model

Now, by following the docs on route creation and handling, the developer can create the necessary endpoints to support the requirements of the project that they are building. For example, they can begin by using the file-based router to define two endpoints – one to create new *package* entries and another to list them. To do this, they can make a directory called `api` and add a new file, `api/packages.js`. You can find an example of this file at `chapter7/componium/tests/sample/models/package.js`.

Instead of doing this manually, the scaffolding tool can also help generate a new route file and a new model file, later placing them into the correct directory. For the route generation, the command looks like this:

```
> componium create
? What would you like to scaffold? Route
? Enter a name for your ROUTE packages
```

This will now provide an endpoint to handle requests. The developer will likely start looking for ways to enrich the API endpoints with actual data. The API creation step requires the application author to understand the available options and the mechanisms to create new endpoints. Most likely, they can learn this from the framework documentation and provided sample apps.

Expanding the API

With the API routes functioning, it is time to enable data persistence. In this case, the developer needs to add database capabilities to save and list *packages*. The framework already ensures that it can connect to a development database locally so that step is taken care of. The next step in the workflow is to add a *Package* model and load it in the API endpoints. The scaffolding tooling can generate a model file in the right location using a CLI prompt or command. For example, the CLI can run componium create to prompt the developer for the database model details. To successfully achieve this workflow task, the application author needs to be aware of the scaffolding tools or manual ways to manage database models within the framework. Once we create the model, we can update the models/package. js file to store the properties of different packages:

```
import Package from "../../models/package.js";

export default async (req, res) => {
  const model = await Package();
  const sample = await model.create({
    title: "Paper Delivery",
    address: "123 Main St.",
    created: new Date(2023, 1, 1),
  });

  const packages = await model.findAll();
  componium.logger.info(`Found packages ${JSON.
    stringify(packages)}`);
  res.json(packages);
};
```

In the preceding code, we can both create new packages and then respond with all the packages in the database. Furthermore, we can split the creation logic and the query logic into different routes. The full packages file can be found in tests/sample/routes/api/packages.js.

By the end of this workflow step, the endpoint should now have the code logic to interact with the newly created Package model and list the records when the packages.js route is accessed.

Adding tests

At this point in our workflow, we have a working API that interacts with the database. The developer has already tested the endpoints manually using sample requests. They can also land some functional test cases to make sure the API works properly. To successfully add tests, there needs to be documentation on the suggested ways to test endpoints. These suggestions can include the use of third-party test libraries or the built-in componium-test library from *Chapter 6*.

Configuring environments

With the routes tested, it is time to attempt to deploy the application to see whether it works in the remote environment. The application frameworks that we saw in *Chapter 1* pride themselves on the ease of deployment. Therefore, the expectations are high to make the application as easy as possible to deploy.

To successfully achieve this step, our framework provides a production configuration file, `config/production.json`. This JSON file contains various environment-specific settings that the application uses when running in the production environment. It is still up to the developer to properly understand how to securely specify the database information and other configuration options. The framework documentation can guide the application authors to suggest optimal ways to make this step work. While the framework provides this file, it's still the responsibility of the developer to understand how to securely specify the required attributes. The way these details are specified can significantly impact the security and performance of the application, making it critical to get it right.

Deploying the application

With the production configuration correctly configured, the developer can now deploy the application to their server environment and test out the new API. This step completes our sample workflow, and if the developer can successfully test out their changes and interact with the database, then the workflow is successful.

This is just one example of a framework workflow, which allows us to document the steps from installation to a working application. There are more steps that could potentially be added; it largely depends on how far we are willing to explore this workflow. For example, using the middleware APIs of the framework, we can explore how common middleware such as authentication can be easily added to the new endpoints. We also didn't cover the use cases where an application author needs to have frontend views to manage and interact with the endpoints.

The process of figuring out these types of workflows can help us identify friction and opportunities to make the framework's developer experience much better. It also ensures that we figure out what types of documentation and tooling improvements can be added to the project as we develop it further.

At this point of deployment of the application, we conclude the workflow's standard progression. The additional steps in the more extensive workflow can involve doing more in-depth database operations and using the framework's GraphQL features. Overall, focusing on several of these types of workflows can help framework authors fine-tune how stakeholders interact with their systems. In the following section, we will take a look at a list of the external dependencies that made all this possible.

Dependencies

The workflow that we mentioned in the previous section is made possible by several external libraries and modules. Here's a recap of some of the modules that we used in the Componium framework from this chapter:

- `@apollo/server` and `@graphql-tools/schema`: The combination of these two tools allows us to offer the GraphQL features of this framework project. Apollo Server is able to integrate with Componium servers, and it also provides an easy-to-use sandbox to test GraphQL schemas.

- `Chokidar`: This is the file-watching library that helps to create a better experience by watching for changes to the application files and performing steps, such as restarting the development server.

- `@inquirer` and `yargs`: These libraries allow us to create the `componium` command-line tool. *Inquirer* can create interactive terminal interfaces, useful for Componium development commands, such as `componium create`. *Yargs* helps us work with command-line commands, flags, and options, making it easier to quickly develop a sleek development interface for our project.

- `express` and `body-parser`: These are the underlying server libraries that make it possible to initialize Componium servers and add routes and middleware.

- `Winston`: This is the logging library that is used in the underlying `Logging` class. It helps us provide a way for the Componium applications to log to different types of logs.

- `sequelize`: This is the ORM layer library that helps applications integrate with a variety of databases.

- `componium-test`: This is the testing library from *Chapter 6* that we can utilize to test the backend framework.

- `debug`: This is the logging module used to track down and debug internal issues of a framework during development. As mentioned in *Chapter 6*, it supports scoping a debug level to a particular component by using the `DEBUG=componium:*` environment variable.

- `config`: This is the configuration manager module that helps store and organize the application configuration in different formats.

Some of these modules are quite common in server-side frameworks and other Node.js tools at large. For the purposes of your own framework, you can choose the packages we just discussed or find alternatives that better fit your use cases. Luckily, the Node.js ecosystem has a lot to offer in terms of ORM, logging, and testing solutions.

Summary

In this chapter, we have taken a step further, from the earlier experience with the testing framework, by composing a brand-new server-side framework that is capable of routing requests, handling API calls, and much more. This supports our plan to develop a full stack framework that covers both frontend and backend features, with components interacting with each other within the same unified vision. Our goal was to create something that is used and reused for multiple application use cases and feature set combinations.

We started by defining our project's goals, and we later developed the core architecture aspects of the framework. This architecture included producing features such as server process management, environment configuration, and database interaction. To enable usability and empower developer productivity, we also focused on producing several features that focus on developer experience.

This was the second practical exercise in our framework experience, and hopefully, this gives you even more confidence in your skills to develop your own frameworks. To take this a step further, the next chapter will focus on our final challenge – building frontend components for our full stack framework. The introduction of the frontend components in the next chapter will enable the whole full stack experience of our newly created framework.

8

Architecting Frontend Frameworks

In this chapter, we now switch focus to the frontend components of the full stack framework that we began building in *Chapter 7*. This is the final part of adding new features and architecting the technical design for the purposes of our **Componium** framework example. The frontend features are the most complex to design because they require a lot of domain knowledge of browsers, in-depth JavaScript, the ability to handle complex edge cases, and so on. We will cover a series of frontend topics that focus on enabling a full stack framework development environment. Here are some of the topics that we will cover:

- **Frontend features**: We will determine the features and goals of the frontend components for our framework. In addition, this new frontend infrastructure needs to interact with the existing components of the full stack framework, such as the backend API routes and the testing interfaces.

- **Architectural design**: After learning more about the low-level interfaces of established frameworks, we will create a framework design that can offer similar features with low-level interfaces of its own. This includes developing a component, view, and routing architecture to server content to the web browser.

- **Frontend patterns**: Learning about common frontend patterns and optimizations will help us become more accustomed to working with existing frameworks and building new ones in the future.

It is crucial to keep in mind that we can only scratch the surface of the feature set that other frameworks offer out of the box. For example, we can include a client-side router with a component-based architecture, including reactivity in those components. Frameworks such as Vue.js, Angular, and Svelte, plus libraries such as React, required years of development to cover all edge cases and significantly expand the feature set. For the purposes of this chapter, we will focus on keeping things closer to the basics and build several technical parts from scratch. This should give you a good understanding of the underlying components of other full stack frameworks with frontend features, such as Next.js. For example, we will use some of the web components APIs that are built right into modern web browsers to enable a rich feature set in our own framework.

Toward the end of the chapter, we shall also examine the intended workflow to understand the series of steps taken by a developer to utilize the framework interfaces to achieve specific application development goals, using our newly architected features.

Technical requirements

The technical requirements for this chapter are very similar to those for *Chapter 7*. This chapter reuses the framework files that we saw in *Chapter 7* with frontend components and interfaces. The sample application in the `tests` directory is also changed to showcase some of the frontend features. Locate the book's repository at `https://github.com/PacktPublishing/Building-Your-Own-JavaScript-Framework`, and continue to use the code in the `chapter7` directory.

Follow the `README.md` instructions in the directory for available scripts. The sample app can be found in the `tests/sample` directory of the framework. When you start that application, it will be available on port `9000`. You can open it with a browser using the `http://localhost:9000/` URL. While you are jumping into exploring the chapter code base, it is recommended to use the debugging tools to trace how the elements come together. Refer to the *Debugging* section of *Chapter 2* to configure a debuggable environment.

Defining the frontend framework

We will continue from the previous chapter by reusing the example **Componium** framework project. In *Chapter 7*, we created several features that allow us to interact with server-side routes, define APIs, and query a database. Currently, there is no way to develop frontend components to either consume those APIs or add visual interfaces using our framework. Without the frontend part of our framework, a developer needing to build an interactive interface, hosted using a Componium server, would need to include an external library and statically serve additional application files from the server.

Therefore, we will change the lack of frontend features by creating several frontend features that will allow framework users to create client interfaces. These frontend features will mimic some of the complex features of the existing established frontend frameworks. For instance, our approach to reactivity features inside the components will include the basics, utilizing the built-in APIs from the browsers.

To begin this process, we will identify the goals of the features we want to support. After that, we will follow with a sample architecture design that will make those features a reality and make our framework genuinely full stack.

Goals

The frontend framework sets three goals that will be later supported by the features that we create. Overall, it is a good idea to define these goals in a general manner and use them as you progress with your framework project:

- **Web-based interfaces**: Goal number one is to empower developers to *create web-based interfaces* while maintaining cohesion with the backend/server components of the framework. With the frontend capabilities of the framework, developers will have the ability to write, host, and deploy interactive client interfaces. These interfaces will be made possible by providing a frontend framework API to create components and attach those to client-based views. These client-based features should also be testable and debuggable, either by the built-in Componium test interfaces or external testing tools.

- **Enable interactivity**: A comprehensive set of APIs, both on the server-side and frontend side, will help *enable the interactivity* required by many of the framework-backed projects. The interactivity features need to enable developers to use familiar technologies such as HTML, CSS, and frontend JavaScript to craft components. The framework should also have the affordances to be able to include potential external libraries. For example, if someone wants to create a visualization within their Componium frontend application, then they should easily be able to include external libraries such as Three.js and D3.js.

- **Promote reusability**: We want to include a set of framework affordances to build complex applications. These could be applications with many frontend differentiated views that include a large number of nested components. These could also include a set of optimizations for production environments and the ability to manage large application code bases. Primarily, we also want to *promote the reusability* of code and guide developers to make intelligent decisions when building their applications. An easily extensible feature set can be beneficial to cover a lot of the potential use cases. If we get the architecture right, it will allow for high degrees of customizability.

Learning from already existing frameworks such as Next.js, we also want to make sure to include some of the more modern features and offer pleasing developer experiences alongside those features. These could include code generation affordances, similar to the ones we saw in *Chapter 7*. To set the framework apart from some of the other solutions, we will also set a goal to use some of the newly shipped Web APIs. Taking advantage of starting from scratch in our project gives us a good opportunity to evaluate the latest developments in the browser platforms, choosing the newly available APIs. As part of our learning goals, we will also try to contrast the difference between our newly developed framework and the established mature frameworks, such as Vue.js and Next.js.

With these goals in mind, let's dive into the set of features that can back these goals up.

Features

The frontend goals defined in the previous section will help us guide the thought process behind our feature development. To support the established goals, let us plan out some of the technical features that developers will find useful and expect from the framework. Here are some substantial features that we will discuss in detail in this chapter:

- **Serve HTML files and content**: To support the features of the interactive interfaces, we will need to add the ability to serve the generated HTML output to browser requests. We need to ensure that we have the capabilities to serve static content to sustain additional JavaScript code, images, media, and other types of files. This feature is vital to render content on the client side in a web browser.

- **Structured application code**: We need to grant the ability to define reusable interactive JavaScript components with CSS/HTML templating and styling features. We will do this using a component-based architectural approach. The component architecture feature itself will enable the development of user interfaces. The structured application code that it can help produce will consist of independent and reusable bits of code that serve as the building blocks of the whole web application. This will support our goals of reusability and providing good application primitives, as it takes advantage of component paradigms, such as reactivity, composability, and modularity. The approach to this feature aims to have each component control its own state and render the interfaces based on the state.

  Supporting the maintenance goals, the component-based approach ensures isolated testing and debuggability of the application code. Here, we channel some of the previous design decisions from other frameworks. For example, in Vue.js, components are structured with JavaScript logic, HTML templates, and the ability to style them with CSS.

- **Composability**: Expanding on the previous point, it is important to highlight the composability of application code as an important feature of the framework we want to create. Defining a nested structure based on how the web browser internal DOM structure, components can then be composed together to create complex user interfaces. Refer to the *Components* section of *Chapter 2* for a recap of the importance of this feature.

- **Client-side router**: One of the core pieces of frontend framework organization that we have seen in *Chapter 2* was the router. The routing feature set is crucial to include in our feature set because it is responsible for navigating between complex views of our application. From *Chapter 7*, we already have an API for server-side routes, and the frontend router will help provide the functionality to enable a fast and smooth transition between different application states.

- The router will adhere to the existing concepts of web application navigation and utilize the relevant Web APIs to modify the browser's URL and history features. The inclusion of the router will also be beneficial in creating a more logically organized code.

- **Interaction with APIs**: The ability to interact with backend APIs is another key feature. The server-side part of Componium has different ways to define endpoints, including the **GraphQL** endpoints. The frontend code should make it easier to make requests to these endpoints and handle the data. In the case of frameworks such as Angular, the project provides the `HttpClient` helper module (`angular.io/guide/understanding-communicating-with-http`) to communicate with suitable backend services. This would be an excellent reinforcing feature for our sample project, especially if combined with some of the specific *Componium server-*defined routes. For example, to help enhance the developer experience, we can pre-generate some of the data fetching calls for the known endpoints and create dynamic interfaces around those, thus saving time for developers building with our framework.

- **Server-side rendering (SSR)**: We mentioned SSR features in other framework examples throughout the book. We will include these features in our example framework as well. The SSR methods will render the components on the server side to help improve the rendering performance of the applications. Given our full control over the developer experience of the full stack framework in this particular case, developing this type of functionality is easier for us. Besides the performance improvements, SSR is beneficial for search engine optimization purposes and general page loading time.

  The internals of SSR include close collaboration between components pre-rendered on the backend and later hydrated by the frontend code. The backend routes should also be able to inject the state of components into the pre-rendered elements. The state can be static data or information fetched from external sources, such as a database.

- **Production optimizations**: As part of the commitment to empowering developers, the framework will also feature some optimization steps for applications running in production environments. This means including additional internal tooling that performs optimizations, such as minification, behind the scenes. These kinds of optimizations are also much easier to integrate within our framework because we have control of the server tooling.

  Besides code minification, we can also look into advanced JavaScript optimization techniques such as tree-shaking and code-splitting. Supporting static file handling, we can potentially optimize other media such as images. Generally, as the framework continues, we want to consistently work on such optimization improvements because it benefits all the framework's users.

The preceding list is a selected set of functionalities that should make the frontend feature-rich, give us a good learning opportunity, and also cover realistic use cases.

The following *Figure 8.1* provides a summary of how the listed features interact with each other:

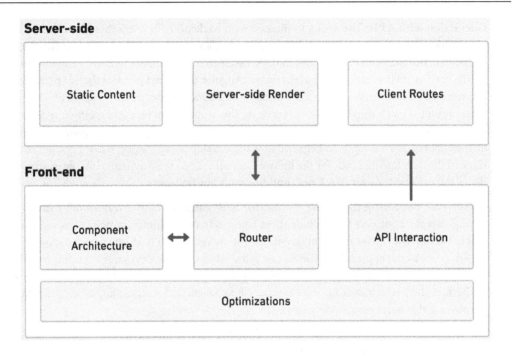

Figure 8.1: The features summarized

The existing server-side code will be able to serve static files that can be consumed by the frontend. At the same time, the server process is capable of importing and accessing some of the components to render them on the server side. Finally, the backend has a definition of client routes; these are the frontend endpoints that are classified as accessible by a browser to be rendered in the client.

On the frontend side, we have the *API interaction* features that communicate with the server or make requests to external APIs that are hosted on external services. Simultaneously, a client-side router tightly works with the component architecture to enable the user interface experience in the browser. Finally, we have a set of frontend optimizations that cover all of the frontend surfaces, ensuring the most optimized experience when deploying and running applications in production environments.

With these features in mind, let us proceed to the architecture step, where we can explore the technical and organizational concepts that make these features possible.

Architecture

With the required features outlined and documented, let us extend the existing architecture from *Chapter 7*. The changes will be adding new functionality to the framework architecture, concentrating on enabling the ClientView abstractions and functionality, which will drive the feature experience behind the frontend changes.

We already have the capability to create server API endpoints. The general implementation of the architecture of the features will consist of introducing several new interfaces to the server part of the framework. The newly added frontend features will be situated in the frontend directory of the framework project.

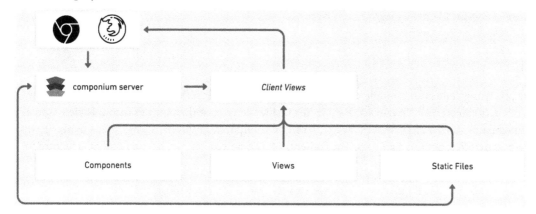

Figure 8.2: Componium frontend components

In *Figure 8.2*, we outline an incoming request to the server powered by the Componium server framework. Specifically, this request is expected to respond with an HTML page to provide interactivity features. This is unlike the API requests from *Chapter 7*, where we would receive JSON or GraphQL responses with data. The request handler can still process incoming request objects, such that it can access properties of the request, such as the query parameters. It can also tweak any properties of the response object. In *Chapter 7*, we used server.addRoute(...) to add new routes. To add route handlers that provide the functionality, we will use a similarly structured server.addClientView(...) method that will have a similar API, but a totally different behavior. This new method is where the **client view** functionality from *Figure 8.2* will take place.

In the Componium frontend component design, a client view can have a singular view and many components used within it. The concept of the client view is a server-side definition, while the concept of views is shared across both the server and the client side. Once the client view is defined, it assembles all the imported components and the View together and sends a response back to the browser.

Besides the client view interaction, the server can now also define and access static file directories. These static files can be directly accessed by the browser, and the whole directory is exposed to the web server. These static files are also usable by the Client Views to import additional resources into the Client Views, such as any media files (images, fonts, styles, etc.) or additional JavaScript components. The ease of access to static files simplifies how a frontend framework can include external media and other useful entities that can be included within the web application.

In the next section, we will get a detailed look into how Client Views become the gateway for the frontend features and files, allowing us to create multiple endpoints serving HTML, CSS, and JavaScript code.

Entry points

To allow flexibility when creating multiple frontend Client Views, our framework provides a way to define multiple client endpoints:

```
server.addClientView("/", "frameworks", {
  title: "Frameworks",
  // …
});
```

The preceding code is an example of routing the root path of our server to the `frameworks` view. Developers can create a `frameworks.js` file inside the `views` directory to map this view by the defined name. `.addClientView(...)` can be configured multiple times, with as many views attached to different route handlers. The contents of `frameworks.js` have a familiar structure to the route handlers from *Chapter 7*:

```
export default function frameworks(request, response) {
  return `<p>Welcome!</p> `;
}
```

The preceding code can be as simple as returning a paragraph tag. You can find other examples of more involved views in the `tests/sample/views/` directory. The view handler has access to the `Request` and `Response` objects of the route to tweak the behavior of the route or fetch additional data. We also have access to the `componium` variable here to access the interfaces of the framework. For simplicity, we will use JavaScript template literals in more complex templates. The framework takes care of rendering the desired HTML structures, and it also wraps the `<p>` tag in the preceding code in a valid HTML document.

The other important entry point for our frontend files is static file configuration. To be able to resolve other types of media, the framework provides a way to mark certain directories of the application project as an endpoint that serves static files:

```
server.addStaticDirectory("/images", "img"));
```

The `.addStaticDirectory(...)` method maps the server route for `/images` to the `img` directory in the application project. This relies on the similar properties of the backend express server that we worked with in *Chapter 7*. This new static directory also works in custom routers, which are created with `server.createRouter(...)`.

We now have a way to process and render basic views and create as many of them as we want. We now need to enable the component-based architecture. This is what will allow us to produce more complex interactive components.

Reactivity

To achieve the dynamic interface features of our frontend framework, we need to learn more about the concepts of reactivity and reactive components. The reactivity concepts in application programming direct the user interface to update and respond to underlying data or state changes dynamically. In the context of JavaScript, especially in the frontend systems, we use custom-developed primitives in combination with browser APIs to achieve performant user interface reactivity. The reactivity features enable a seamless experience for web application consumers by automatically updating the interface whenever relevant underlying data changes. Similar to other programming environments, in the JavaScript world, developers rely on frameworks and helper libraries to enable reactivity for components in their applications. In fact, this reliance on external tooling in JavaScript is much stronger than in other languages. This is mainly due to the cross-browser and cross-engine nature of web page and web application development.

The reactive programming paradigm has become very fitting in the web development environment, due to the asynchronous nature of JavaScript. Other contributing factors for the reactive paradigm fit included application requirements for real-time updates, complex interactivity, and making it easier to manage the state of applications. The user interfaces have become much more complex. These days, the expectations of web applications require frontend systems to be dynamic, consume real-time data, and be instantly responsive to user actions. Also, the existing structure and abstractions around a web browser's **Document Object Model (DOM)** have compelled solutions to be based on manipulating a nested node tree of page elements. The reactive changes in the interface utilize the diffing algorithms to update the changed nodes of a component tree.

Reactivity requires data binding, which is a connection between data and the elements of applications. In the web development world, the data would be provided by JavaScript interfaces, most likely dynamically loaded from some API endpoint. The elements would be the *HTML/DOM* structures in the browser client. The corresponding browser elements are automatically updated to reflect the changes when the underlying data updates. In *Chapter 2*, in the *Framework building blocks* section, we highlighted the possible ways in which data binding can occur within these reactive components. We also saw examples of frameworks using one-way or two-way binding. The flow largely depends on architectural decisions, either allowing the elements to update when data changes or also allowing the elements to update the underlying data. Popular frameworks such as Vue.js and anything that includes React use an implementation of a *virtual DOM* tree to render an application's state as the data changes. However, there are also examples where projects, such as those involving Svelte and Angular, use the real DOM or *shadow DOM* features to achieve similar functionality.

Similar to other frontend projects, we will introduce the concept of reactive components into our framework. These components will allow us to encapsulate HTML elements and application logic alongside them. The frontend *Componium* components will maintain their internal state and respond to data and interactions. To keep things simpler and without diving deeper into the internals of the existing frontend frameworks, we can build an example of basic reactivity concepts using a combination of Web Components and other more modern Web APIs. A practical overview of the component architecture will provide a good learning opportunity to understand the built-in browser primitives. It will also offer a good comparison and understanding of the complex problems that the existing frameworks solve for us:

```
class ReactiveComponent extends HTMLElement {
  constructor() {
    super();
    this.state = this.reactive(this.data());
    this.attachShadow({ mode: "open" });
  }
  reactive(data) {
    return new Proxy(data, {
      set: (target, key, value) => {
        target[key] = value;
        this.update();
        return true;
      },
    });
  }
  callback() { this.render(); this.update(); }
  render() { this.shadowRoot.innerHTML = this.template(); }
  // methods that child components will override
  update() {}
  data() { return {}; }
  template() { return ""; }
}
```

The preceding code is our newly defined ReactiveComponent; it has been compacted to fit in better in this chapter. The class starts off extending HTMLElement. This interface will help us represent HTML elements and create our own web components. In the constructor method, we have a declaration for the state property, which will keep track of the state of the component. Another important constructor call is .attachShadow(). This call attaches a Shadow DOM to the custom element and provides a scoped environment for the encapsulated CSS and JavaScript instructions.

> **Additional reading**
>
> For a detailed low-level explanation of how Shadow DOM is structured, check out the MDN page at developer.mozilla.org/en-US/docs/Web/API/Web_components/Using_shadow_DOM.

In the reactive method, we configure a new Proxy object, which is another built-in Web API with properties that can help us make reactive changes to the state of our components. The Proxy methods take an object and return a new object that serves as a proxy for the original definition. The proxy behavior helps trigger updates and re-renders of a component when the object is updated.

> **Additional reading**
>
> For details on the Proxy interface, check out the detailed article at MDN: `developer.` `mozilla.org/en-US/docs/Web/JavaScript/Reference/Global_Objects/` `Proxy`

The setter operation and the update methods will be invoked when the state changes. The `update()` method is overridden in the components that inherit this `ReactiveComponent` class. With `ReactiveComponent`, we can build a simple set of example components. This structure will bind the data from the component state into the rendered templates.

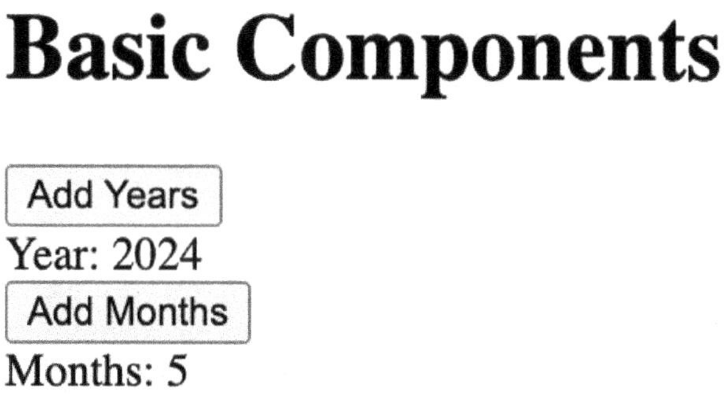

Basic Components

Figure 8.3: A basic components example

Figure 8.3 presents an example view of two components in action; one is the year tracker and the other is the month tracker. Both components have the option to increment the values of the appropriate dates. You can find the code for this working example in the `tests/basic-components/` `index.html` file:

```
<body>
   <h1>Basic Components</h1>
   <year-component></year-component>
</body>
```

The source of the view follows; it just includes a newly defined component by its name, `year-component`, wrapped in regular HTML tags. This was achieved by registering the custom element using the built-in Web Components interface – `customElements.define("year-component", YearComponent);`. The `year-component` component extends the preceding `ReactiveComponent` class and overrides the empty `update` and `template` methods, as follows:

```
template() {
    return `
      <button id="addYears">Add Years</button>
      <div id="yearCount">Year: 0</div>
      <month-component></month-component>
```

```
        `;
    }
    connectedCallback() {
      super.connectedCallback();
      this.shadowRoot.querySelector
        ("#addYears").addEventListener("click", () => {
        this.state.yearCount++;
      });
    }
    update() {
      this.shadowRoot.querySelector(
        "#yearCount"
      ).textContent = `Year: ${this.state.yearCount}`;
```

The methods listed in the preceding code, such as `update()` and `template()`, render the data relevant to the component and define the template returned by the `year-component` component. We also have event handlers that change and update the year, using access to `this.state`. Also, note that to access the Shadow DOM properties of this `Year` component, we use `this.shadowRoot.querySelector`. The template defined in the component includes `month-component`, which is a nested component. It has a similar setup to the extended class of `ReactiveComponent`.

This component configuration, using the Web Components APIs and other affordances from the browser APIs, is a good starting point for the framework. We can use these patterns to achieve similar functionality in other frameworks, such as Vue.js, and frameworks that use React as their underlying library to structure their components. The interface we have has reactivity properties, the ability to compose components simultaneously, and the basics of templating.

In the following section, we will take this a bit further and utilize an external Web Component helper library to build upon this pattern.

Improving components

In the reactivity part of the architecture, we have described a pattern of using Web Components comprised of other techniques to achieve the desired feature set. To enhance this further, we will bring in the **Lit library** (`lit.dev`), which will help make our job of managing the components much more straightforward, and our framework can be dependent on it. We shall use the same approach of abstractions and utilize the interfaces of this library to create the component features of our framework.

About the Lit library

The Lit open source library has been around for more than six years, and its goal is to simplify and abstract away some of the verbose tasks that deal with Web Components. At its core, it provides reactive state features, scoped CSS styling, templating, and a variety of advanced features to compose and work with frontend components. It works with both JavaScript and TypeScript languages and comes with a large variety of packages that extend its functionality. For example, Lit has additional tools to enable localization, animate frontend elements, and also integrate with React components.

The library uses the standard component life cycle, in addition to its own component life cycle methods to simplify certain operations, such as DOM reactivity. You can find the source of the project at `github.com/lit/lit` and `lit.dev/docs`.

To draw a comparison with the preceding `ReactiveComponent` interfaces, let's take a look at how similar the code would be to the basic components if we used the Lit library:

```
import {html, css, LitElement} from 'lit';
export class YearComponent extends LitElement {
    static styles = css`p { color: green }`;
    static properties = {
        year: {type: Number},
    };
    constructor() {
        super();
        this.year = 2024;
    }
    render() {
        return html`<button id="addYears">Add Years
          </button>
      <div id="yearCount">Year: ${this.year}</div>`;
}
```

The preceding code imports the `LitElement` class that we can extend. This new code block result looks very familiar to the code of `year-component` that we saw in the *Reactivity* section. However, there are some additional improvements we have in this component definition.

Let's explore a few of them:

- First of all, we have a CSS helper interface, which allows us to declare the style of our component using the `static styles` variable.

- Second, the way we declare the state of the component has also changed – we define a `static properties` object with some extra definitions of the properties. These properties are used in a similar way in the template.

- This brings us to the third point – the templating is also a bit different. It uses the `lit-html` helpers to enable more advanced templating features and help us work with HTML. This helper allows us to create templating directives, tweak the rendering methods, and so on.

> **Additional reading**
>
> Detailed documentation of all the templating features can be found at `lit.dev/docs/v3/templates/expressions`.

For the purposes of our sample framework, we will define our own `ComponiumComponent` class. This class will be available to the developers to create rich components, combining what we have learned from the *Reactivity* section and the Lit library. We can also rely on the rich functionalities of Lit templating to render the results. To enable this, we will load the Lit library alongside the `componium.js` framework file in `ClientViews`. This will expose the component interface to developer-defined components. To start using those interfaces, developers can import them using the ES6 syntax:

```
import { css, html, ComponiumComponent } from "componium";
```

The class to extend is provided, alongside the CSS and HTML helpers to help construct the components. For example, if our component has interactive buttons, it can use the following Lit syntax to declare event handlers:

```
html`<button @click="${this._click}">click</button>`;
```

The `_click` event handler is a method defined on the class that extends from `ComponiumComponent`. If these components require any static files, they can request them directly by fetching them from the static routes declared by the Componium server. However, we can still take this one step further and use Lit and the mix of our framework's interfaces to enable the complex features of SSR. There will be more on the concept of utilizing the existing components from a server in the next section.

SSR

We learned about the features and benefits of SSR when we planned the frontend interfaces of Componium. On a technical level, SSR requires several parts of our framework to work really well. Here are some of them:

- The component architecture needs to support different rendering capabilities. This includes the ability to pre-render the components as HTML that can be transmitted over the wire to the frontend.

- Components need to be able to run consistently on both the server and the browser.

- The server-rendered components should have the ability to fetch data in both the client and server environments. Depending on the environment, components should have a similar approach to how data is requested and processed.

- We need a server that can structure and render the components, including attaching a state and data to these components.

- The frontend side should be able to take the server-side state of the components and later hydrate them. It should be able to attach the required events.

Luckily, with our framework and the Lib library, we have a lot of these requirements covered and can develop the SSR support in our framework. Using the `@lit-labs/ssr` package, we can define a server-side render in conjunction with our Client View abstractions. The modules for these features can be found in `packages/frontend/client-view.js`. To have a flexible feature set in our framework, we want developers to use both client-side-only components and a combination of server-rendered components.

To enable the SSR capabilities, the framework has a new `Renderer` class, which has the job of consolidating all the required framework code and developer-defined components. It does so by responding to client-side requests with a unified template of a valid HTML document and by injecting code from the application directory. To render these structures, we can use the `html` helpers from the `ssr` library. To understand the capabilities of the `ssr` package, check out the comprehensive documentation at `lit.dev/docs/ssr/overview`.

These capabilities work in tandem with the Componium server rendering methods to output the resulting HTML to the client. Once the HTML document has been fully loaded in the browser, then the hydration process begins. The framework files load the required supporting files that will help attach the event handlers to our components and make them interactive.

We shall examine the detailed usage of the server-side rendering features in the upcoming *Using the framework* section. Meanwhile, there are a few more architectural features left to cover before we can start using all of them together.

Client router

At the framework planning stage, the router was highlighted as an essential piece of frontend infrastructure that allows an interface to transition between significant sections of an application. The routing implementations are very similar across many frontend frameworks. However, if you are really passionate about the routing features, the *React Router* project (`reactrouter.com`) is a great project to learn from in terms of educating yourself about good routing abstractions, the potential pitfalls, and routing edge cases.

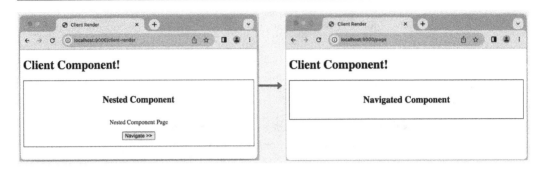

Figure 8.4: A transition between the initial page (left) and the newly routed page (right)

To enable the routing functionality in our framework, we will introduce a ComponiumRouter class. In *Figure 8.4*, we can see the routing transition between the initial /client-render page (*left*) and the newly routed page, /page (*right*). The routing happens when you click on the **Navigate >>** button in the component. The routing interface can be initialized within our components and allow an application to navigate between the nested component templates. The public developer API for the router is this.router = new ComponiumRouter("client-render", routes, this.shadowRoot);.

The router gets an identifier (client-render), a series of routes with the routes object, and a root element (in this case, the this.shadowRoot object) that will be used to render the routed templates. The routes object is defined as an array of objects. For example, a simple route example for *Figure 8.4* would look as follows:

```
const routes = [{
    path: "/client-render",
    template: "<nested-component></nested-component>",
  },
  {
    path: "/page",
    template: "<navigated-component>
      </navigated-component>",
  },];
```

The /client-render route is one of our entry points that has Client View type and the /page route is a view we can navigate to. The router can support a richer feature set by accepting more complex templating structures in the template property. For example, we can use the html helper from the Componium/Lit modules to produce more extensive template objects in the routing definition.

To navigate between routes, we will rely on more built-in Web APIs in browsers. For instance, the components call out to history.pushState(...) when views need to be changed. Inside the ComponiumRouter class, the framework handles those pushState events and renders the appropriate template.

> **Additional reading**
>
> The MDN documentation outlines all possible History API methods we can use in our frontend routing components. It is available here: `developer.mozilla.org/en-US/docs/Web/API/History/pushState`.
>
> In the Lit Labs open source code base, there is another example of a component router. The source for it can be found at `github.com/lit/lit/tree/main/packages/labs/router`. It can be an excellent exercise to implement the router in the *Componium* framework.

Optimizations

Part of our goals for frontend functionality was a set of optimizations to make the applications that our framework produces more efficient, scalable, and performant. We shall take a step toward that by introducing some features to optimize the output of our frontend components in production environments.

We will introduce a new `Optimize` class, which has some functions to perform optimizations on the code base. The class can be found in the framework directory at `packages/frontend/optimize.js`. These optimizations can affect both the included/injected framework files and the application code. The functions of this class will activate when applications deploy within an envrionment that has the `NODE_ENV=production` variables defined, which is the common pattern for Node.js-based projects.

We will utilize some existing JavaScript tooling – in this particular case, `esbuild` – to `minify` the framework and component code files. The `esbuild` tooling provides the following minification command-line API:

```
esbuild ${filePath} --minify --outfile=${newFilePath}.
```

We can use the power of `esbuild` tooling to optimize applications built with our framework. For example, all the components used in an application will be minified by the server process when required. Under the hood, the framework parses through the component files and runs the minification step, outputting the optimized files into a separate directory. We use a hidden directory called `.componium` as storage to save the optimized files. The framework later knows to access the optimized files instead of the originals.

To expand further improvements in the future, we can focus optimizations on other application files, such as images, media, and more. The introduction of more complex build tools is also possible. For example, we can add `rollup.js` to enhance the resulting output of the client-side code. We have seen example usages of rollup tooling in *Chapter 3*. `Esbuild` also provides additional functionality besides the minification that can be found at `esbuild.github.io/api`.

Now, the final part to cover is the improvements in the developer experience for these frontend components, which we will do in the following section.

Developer experience

The finishing touches for our features are to include some improvements to the developer experience. We will do this through thorough documentation of the component system and by providing enhancements to the framework's executable file.

The documentation should provide clear instructions on the frontend capabilities, such as the definition of multiple Client Views and static file directories. The component structure, composability, and reactivity features also need to be described. For example, this could include a straightforward API to add new client routes and how the framework uses the Lit library to achieve the component-related features.

The improvements of the framework executable include the ability to generate the Client View routes, Views, and Components using a scaffolding operation. In *Chapter 7*, we saw an example of generating new API routes; this is a very similar addition. Just as before, with the executable, developers will be able to quickly generate some code and start composing frontend user interfaces. These pre-generated components include the default property configurations for styles and data properties of objects.

Overall, we will demystify as many of the frontend features as possible, focusing on helping developers navigate the complicated intricacies of all the technologies that are in play while building client-side web applications.

Dependencies

We will use several dependencies fetched from the npm package registry to achieve the level of functionality described in this chapter. Here are some notable ones:

- Esbuild: The bundler and minifier tool used for optimization steps when the framework server runs in production environments. With esbuild, we can quickly optimize scripts. It includes a lot of advanced features that we can use to extend the end-result scripts of our framework further.

- Lit: The library that helps us extend the existing Web Component techniques and provides much more advanced component features, such as enabling easier data-binding and simplified state management.

- lit-html: Another module that is related to the Lit library, which provides templating features for the frontend features of our framework.

- @lit-labs/srr and @lit-labs/ssr-client: These two modules enable the SSR features of our framework. They can render the components in the Componium server and are later hydrated on the frontend.

Mainly, these libraries and tools help us enrich our framework features. Our framework can rely on these dependencies to enable efficient project building, component-based architecture, dynamic content rendering, and SSR, resulting in a performant, maintainable, and user-friendly application. With the detailed architecture in place, in the following section, we will explore the developer workflow to create a simple client-side application with Componium's frontend features.

Using the framework

Now that we have the architecture in place, we can go through a scenario where a developer uses our frontend framework parts to build a simple frontend example application. In *Chapter 7*, we performed the same task to get a good outline of all the features working together to achieve a certain task. To follow along, ensure that you have installed the dependencies, and then you can start the sample app in the following directory:

```
> cd chapter7/componium/tests/sample
> componium dev
Executing (default): SELECT 1+1 AS result
Componium Server (PID: 59938) started on port: 9000
```

You will be able to open the browser at http://localhost:9000 to view the application.

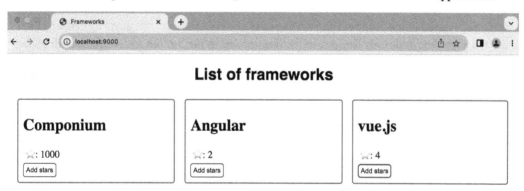

Figure 8.5: A sample client-side app

Based on *Figure 8.4*, we want to create a page with a header called **List of frameworks**. We also want to list some frameworks on this page. Every framework entry should attach to the client-side view and show a number of star points to every framework. Besides the static functionality, we have buttons under each framework component that are able to increment a counter, in this case we label the counter as the number of stars for every project. The developer use case also includes using the SSR features to render the page from the server for performance and SEO purposes.

To begin, our Componium framework allows us to create ClientViews from the server. We can create a client view at the root of the application in the app.js file. We can also utilize the componium executable here to scaffold the components:

```
server.addClientView("/", "frameworks", {
  title: "Frameworks",
});
```

The following code from the root of our application will establish a hosted route at the endpoint designated with a / on the Componium server. Besides providing the path, we also specify the name of the view and additional options for that view. The name of the view maps to the `views` directory in our applications. The `views/frameworks.js` file defines our server-side component handler:

```
export default function frameworks(request, response) {
  const hitsComponium = 1000;
  // code omitted
  return `
    <div class="framework-list">
      <framework-item name="Componium" count=
        "${hitsComponium}"></framework-item>
      <!-- extra code omitted -->`;
}
```

To see the full version of this component, check out the `tests/sample/views/frameworks.js` file. The preceding code only lists one of the frameworks in the list, but the sample has all of the required items. Looking deeper into the code, we have access to the `request` and `response` objects in this exported `frameworks` function. This is where developers can also access the database ORM methods in this file to fetch the data and pre-populate the component state.

With the defined `framework-item` components in place, we can start the application server and navigate to the root of the application. If we view the source of the file, we can see the SSR components. Partially, the source would look like this:

```
<h1 style="font-family: sans-serif; text-align:
  sscenter;">List of frameworks</h1>
    <div class="framework-list">
      <framework-item
        name="Componium"
        count="1000"
      ></framework-item>
```

The state of the component, including the number of stars is hydrated from the rendered component. If we use the following `button` element in the `framework-item` component, then we increase the number of starts based on the state that originated from the server:

```
<button @click=${() => this.count++}>Add stars</button>
```

Finally, with the `frameworks.js` application view created, we can now deploy our application to give it a test run. Here, developers should configure the `app.js` process to run with the NODE_ENV=production environment variable turned on. This will enable the `esbuild` optimization features, allowing the server process to minify our newly created components.

This example use of the framework included creating components, rendering them on the server side, and interacting with their state by clicking on their UI elements. This practical example and the routing examples from the *Client router* section showcase most of the features that we have developed for this initial version of the framework features. The following steps from here could involve finding ways to improve the component-based architecture, as well as finding more ways to add potential optimizations to the applications built with our framework.

Summary

In this chapter, we focused on building a frontend architecture and adding frontend features to our existing project. Similar to the definition of server-side architecture in *Chapter 7*, in this chapter, we had to define the goals behind the frontend features, focusing on what developers would like to do with our full stack framework. We have covered the topics of defining entry points for client routes, concepts of reactivity, complex component structures, SSR, routing, optimizations, and so on. The frontend feature set can be overwhelming, with a lot of terminology, and there is much more to learn beyond this chapter.

If we combine all the components that we have architected in the past several chapters, we now end up with a framework consisting of three use cases that combine into a larger full stack narrative. So far, we have seen a JavaScript testing framework, a backend framework, and finally, a frontend framework under the same logical namespace.

In the next chapter, we will focus on essential topics of framework maintenance by shedding light on various situations that can occur as frameworks evolve.

Part 3:
Maintaining Your Project

In conclusion, the last two chapters give an in-depth look at the maintenance aspects and future of framework projects in the JavaScript programming space. The driving factor behind these chapters is to ensure the longevity and usability of the projects that developers build to guarantee the creations' reliability and effectiveness. Looking over the best practices of such systems, the final chapter presents current and future ideas that are relevant to established and new projects.

In this part, we cover the following chapters:

- *Chapter 9, Framework Maintenance*
- *Chapter 10, Best Practices*

9

Framework Maintenance

In our framework development journey, we have reached the point where we can discuss framework maintenance in detail. So far, we have completed an initial version of the full stack framework in the previous chapters.

In this chapter, the framework maintenance topics will guide you toward future frameworks to develop. These are the topics related to the framework release process, the continuous development cycles, and, finally, the long-term maintenance of large framework projects. In more detail about framework maintenance, we shall learn about the following:

- **Development cycle**: We'll learn about the concepts and paradigms that help develop formal feature definitions and figure out how to seek framework user feedback. First, we shall recap the analysis and design steps of the framework development process. Then, we'll move on to formal feature definitions, as well as ways to seek user feedback and keep these users informed about changes in the framework. Being adept at these concepts will help you keep your framework relevant, user-friendly, and competitive.

- **Release process**: Getting accustomed to the release process involves several logistical tasks, such as defining the versioning and licensing terms, combined with a pipeline that can release and deliver your project to the public. This entails creating automated pipelines to build, test, and deploy your framework. This process also teaches you how to smoothly transition from code completion to a published product that users can easily adopt and benefit from.

- **Long-term maintenance**: This part revolves around broader aspects that extend beyond typical coding tasks. The long-term framework management tasks include monitoring a project's health, keeping it secure, managing dependencies, and making necessary improvements.

Understanding these three aspects – the development cycle, the release process, and long-term maintenance – will significantly enhance the success of the frameworks you build and maintain. It will ensure that your project stays relevant, secure, and robust and is of the highest quality, ultimately leading to a better response from its stakeholders.

Technical requirements

Just like in the other chapters, we will continue to use the repository at `https://github.com/PacktPublishing/Building-Your-Own-JavaScript-Framework`. The `chapter9` directory for this chapter consists of several sample projects that showcase tools related to framework maintenance. Most of these have a similar structure to a common Node.js project. For each child directory in `chapter9`, you can explore the maintenance tool examples. Refer to the `chapter9/README.md` files for extra information about the included projects.

The development cycle

To better understand the framework maintenance tasks, we will rely on the **Software Development Life Cycle** (**SDLC**) to partition some of the important milestones of our framework development, such as building features and releasing them for your stakeholders. Our goal is to take the broad themes of SDLC and apply them to our framework projects, focusing on specific examples that will help you build better projects.

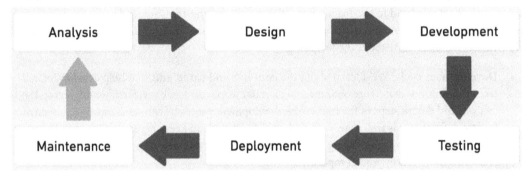

Figure 9.1: The software development life cycle

Figure 9.1 shows a simplified diagram of the SDLC, which we can apply to our framework projects. In this book, we have performed these steps in detail, except we need to focus further on the **Deployment** and **Maintenance** steps to complete the cycle. We have taken care of the first **Analysis** step by learning about the organization of other frameworks and planning out each large portion of the *Componium* full stack framework. We have performed useful examinations of the existing abstractions, popular patterns, and types of frameworks in *Chapters 2 and 3*. To further reinforce this **Analysis** step, we brainstormed with potential stakeholders and the feature set of the testing, backend, and frontend parts of *Componium*. In this chapter, we will look at other strategies that help framework developers make framework projects more successful.

About SDLC

The **Software Development Life Cycle** (**SDLC**), in some cases also called the **Systems Development Life cycle**, is a term that originated in the 1960s. At the time, it was used to help develop complex business systems. Later, it was used further in software projects, combined with different development models such as waterfall. This term defines the systematic process of building large systems and software, providing the methodology for a development process, and focusing on quality and programming efficiency.

Software developers usually use this model as a foundation and evolve their approach to building software, based on what works works better in their organization. The introduction of agile software development methods is an extension of an approach to SDLC. In the instance of JavaScript application development, development teams can use these life cycle concepts to build new projects, with the use of framework tooling and libraries. Simultaneously, framework developers can also rely on the SDLC methodology to craft a high-quality, performant framework experience.

The **Design** stage, following the **Analysis** step, is what we discussed in when building a framework in *Chapters 6*, *7*, and *8*, and it outlines the architecture parts of a framework. This is where framework authors focus on bridging the gap between requirements and the start of writing code. The **Design** step is particularly interesting when working with the JavaScript ecosystem. Due to the multifaceted nature of web development, including different aspects of JavaScript, runtimes, and web APIs, there is an opportunity to create innovative design approaches, combining the existing and newly built components. For example, in *Chapters 6*, *7*, and *8*, we had to make careful decisions about the combination of abstractions and the choice of libraries that we intended to use within the full stack framework. Going beyond the choice of libraries, the elements of the **Design** stage require making concrete decisions about the combination of technologies that will be used in the framework projects and how these technologies communicate together. This is where we strike a balance of strong opinions versus a flexible framework.

However, the whole concept of the SDLC process is a *cycle*, which means the design decisions will return to a developer in a cyclical manner. This conveys that not all decisions are set in stone and can be changed over time as a project progresses. However, some decisions are harder to change than others. For example, changing the API structure in the *Componium* testing framework will require all the consumers of the framework to rewrite or migrate their tests to a newer API. Most likely, the API change is for the benefit of the framework and its users, but it still creates friction for the projects utilizing the framework. This can result in framework users being stuck on certain versions of the project or moving off to use something else entirely.

As a framework developer, you will reach the **Design** stage again after a project's initial release. When that happens, the framework SDLC will be more narrowly focused on feature development and larger refactoring of code. In the real world, an example of this stage would be in the Vue.js project. The major versions of Vue.js introduced changes to the renderer, made it easier to work with template directives, and changed the API of components. More details on that migration can be found at `v3-migration.vuejs.org/breaking-changes`. With cases like these, the framework authors go back to the drawing board or find ways to drastically redesign certain public-facing and underlying structures of a framework.

This cycle of coming back to the **Design** stage is common in the evolution of a framework, particularly when it's in active development or has a thriving user base. While potentially disruptive to the existing users, this iterative process is key to the framework's long-term success and sustainability. Frameworks that fail to adapt or evolve can quickly become outdated and lose relevance. Consider the evolution of the Angular framework project. A complete rewrite was undertaken in its evolution from AngularJS to modern versions of Angular. While initially disruptive to the developers, this transition allowed the framework to modernize its architecture and take advantage of the latest JavaScript features. It also set the foundation for a more future-proof growth path, ensuring Angular's position as one of the top frontend JavaScript frameworks.

Another area where the **Design** stage reoccurs is the introduction of performance improvements, specifically those affecting JavaScript projects. Given the nature of the JavaScript language and ever-evolving web technologies, performance optimization is often an ongoing endeavor. The team behind the React library has consistently revisited its design phase to make performance improvements, leading to significant enhancements such as the introduction of Hooks, concurrent mode, and the new JSX transform. The SDLC **Design** stage keeps you busy, as it is a continuous activity that enables a framework to stay relevant, performant, and useful to its users. The key is handling these design changes in such a way that it minimizes disruptions and provides clear paths for users to adapt to the changes. Continuing on a similar theme, in the following sections, we will learn more about what framework authors can do to safely implement and maintain new features.

Seeking user feedback

One of the most exciting parts of framework development is getting some stakeholders to use what you have developed. These stakeholders could be teams within your company, external open source users, or even your own team. The framework product that you produce needs an efficient feedback loop. This loop will consist of finding ways to get the best feedback you can from as many stakeholders as possible. After successfully acquiring the required input, a framework developer can outline the required changes to adapt to that feedback. These changes can include fixes to issues, the addition of new features, improvement of the developer experience, or overall organizational tweaks. Once these changes are in place, we reach another exciting part of the framework development process – delivering the newly updated version of the framework to the users.

In this part of the chapter, let's focus on the first part of seeking user feedback, which is properly gathering feedback. Depending on the scale of your project and organization, you may end up interacting with large groups of stakeholders or having a one-on-one interaction with just a few users. For larger groups, it is beneficial to utilize a system that can facilitate discussion and the ability to have different discussions related to various aspects of a framework. The most simplistic approach is to maintain an issue or feature request tracking system. This is very similar to other software projects, but with a framework, investing more time in managing the aspects of this system is more important. With a framework, you want to be able to clearly separate the issue and feature discussions, slating the feedback for different releases of the project.

The framework can also provide opportunities to direct feedback to the right place. For instance, gathering developer feedback can be embedded within the developer experience of a project. This is where the executable that developers use daily can provide links to the feature request tracker. In *Chapter 7*, the componium executable had mechanisms to generate scaffolding of certain common components, and the feedback direction could be built right into that executable. Overall, you must rely on tools and systems to gather feedback and avoid spending too much time manually managing all of it. Most importantly, it is important to be respectful of all types of feedback. In some cases, the features built and described within a framework may not work well for some stakeholders. Therefore, it is essential to appreciate each piece of feedback as a unique perspective that brings with it the prospect of innovation and expansion of your project. In the next part, we will discuss a formal feature definition process that worked well across open source and private framework projects.

Formal feature definition

The **Request for Comments** (**RFC**) and **Request for Proposals** (**RFP**) processes are commonly used across many industries to define new features or get feedback. These same processes work really well in software framework development as well. Many frameworks have implemented their own approach to this process of formal feature definition, with stakeholder feedback built in. For example, Ember.js hosts its RFCs at github.com/emberjs/rfcs, and Vue.js hosts RFCs at github.com/vuejs/rfcs, keeping track of all new, pending, and archived proposals for feature additions and drastic changes to the project. With the open source nature of those frameworks, anyone can chime in as new feature proposals arrive in those repositories. The proposals get reviewed in a similar manner to code contributions. Conveniently, these feature definitions are stored in a code repository so that they can be logged and referenced in the future.

Including a process for formal feature definition helps organize both framework developers and users to participate in the **Design** stage of the **Framework Development Lifecycle** (**FDLC**). As a framework developer, you are free to choose how involved this process can be. For example, if we added a new feature to Componium to get native WebSocket support, we could create an RFC explaining the gist of the feature, propose an API for it, and state why it would be beneficial to the full stack framework overall. Depending on the complexity of our organization, the proposal to add this particular feature can transition through the following stages:

- **Proposed**: When the new WebSocket feature is described in full detail. At this stage, other framework developers and potential stakeholders get the opportunity to provide initial feedback. In our example, this is where the public API, frontend and backend integration, and the coverage of the WebSocket feature are proposed.

- **Exploring**: The stage where framework developers get to explore technical prototypes and explore the potential architecture of the feature. During this stage, implementation details can be refined. For a framework project, it is also a good idea to share the feature with the teams that have a stake in the framework to seek further feedback.

- **Accepted**: If the feature is deemed sufficient, then the framework can proceed with the implementation and merge the code into the main code base at this stage.

- **Ready for release**: The pre-release stage is a good opportunity to create a new **release candidate** (**RC**) for the latest improvements to the framework. There is another opportunity to get useful feedback on how the feature works with the existing projects and integrations. In terms of maintenance, user documentation can be introduced.

- **Released**: The final stage is where the feature is launched and made available. This is the point at which the RFC can finally be tagged as complete and released. Future proposals can also reference this feature to help with feedback and the technical architecture.

Having a formal process like this helps structure a well-organized approach to feature development and user feedback. However, it is worth noting that many features can get stuck at the **Proposal** and **Exploration** stages.

A real-world example of a simple RFC that was accepted and merged is the removal of Internet Explorer 11 support in Vue.js: `github.com/vuejs/rfcs/pull/294/files`. The documentation mentions the motivation behind the change, including the concerns about the maintenance burden and how the change affected the consumers of the framework. Over 50 replies were contributed to the discussion thread of that proposal, including members of the framework's core team and other developers passionate about the change.

The RFP process can work in a similar way to trigger interest or a bid from another entity to help contribute to the framework. If there is some feature that you would like to add to your framework but are not able to, then through this process, you can find a vendor that will do it for you. This process is more common in corporate proprietary environments.

Both of these approaches create a structure around feature development and help us follow the SDLC in our own way that fits well with the project. As your framework evolves and develops, you might choose to tweak how these formal feature definition processes perform best for your needs.

With the feature development cycle in place, in the next section, we can now highlight the ways a JavaScript framework approaches the release process, including providing the ability to report to its stakeholders about all the new improvements and features shipping in the new versions.

Release process

In the SDLC diagram in *Figure 9.1*, the **Deployment** step signifies when software can be utilized by consumers. In the context of framework development, this means a new release that is made available. This is where newly crafted features become available, and to make them available, developers need to go through the **release process**. In this part of the chapter, we shall explore the topics related to the initial and subsequent release of JavaScript framework projects. This will include showcasing some of the existing tools, licensing options, versioning, and continuous delivery. To follow the *The development cycle* section, where we discussed the introduction of new features, we will begin by learning how to keep everyone informed of the changes to a framework. Later, the *Simplifying releases* section will discuss the opportunities to make the release chores more approachable.

Change logs

Framework developers already have to spend a lot of time planning, architecting, and developing features. Luckily, for the final step of releasing those features, the process of collecting the details of all the newly built framework components can be standardized and automated. To make this happen, projects rely on existing tooling, such as release utilities and change log generation tools, to support publishing new releases of a framework. The first thing that is useful to maintain and generate is a log of changes in every release. The change log is useful in any framework setting and should be the way you communicate with stakeholders about the impact of every new release.

Here are some good examples of change log structures from other popular projects:

- **Electron**: The application framework project hosts its change log at `releases.electronjs.org/releases/stable`, with very detailed information on changes in stable releases and upcoming pre-releases. It gives you the ability to filter by major framework versions and integrates with landed code by linking directly to the source at GitHub. The calendar of releases at `releases.electronjs.org/history` also provides a great visual of all the newly tagged versions. This example could be overkill for a newly created framework but works well for an established project with a large user base.

- **SvelteKit**: The framework takes a simpler approach to managing the change log. It hosts this log of changes at `github.com/sveltejs/kit/blob/master/packages/kit/CHANGELOG.md`. The log keeps track of the releases and the changes within those releases by category. For instance, types of changes can include *feat* (feature development), *docs* (documentation changes), and *perf* (performance improvements). The log update process is automated using the `changesets` automation tool.

- **Next.js**: The framework uses the *GitHub Releases* feature to outline the changes in `github.com/vercel/next.js/` releases. Using GitHub's interface allows the project to document the changes, including the contributors, assets, and source code on the same page. GitHub's tooling allows you to run comparisons across releases to get a series of commits that have changed from a previous release. The creation of the change log itself is done with manual scripts within the framework at `github.com/vercel/next.js/blob/canary/release.js`. These scripts are automated using *GitHub Actions* when triggered by framework maintainers.

The common theme around these change logs is that, often, the change logs are automatically generated, based on certain code commit message structures and with the use of different automated tools, which are published for users to browse through. Depending on your requirements, you can benefit from the following tools or similar in your JavaScript framework projects:

- **commitizen** (`commitizen.github.io/cz-cli`): A command-line tool specifically to make the process of writing and logging commit messages organized. If you integrate this or a similar tool into your code workflow, the whole team contributing code to the framework will produce consistent and quality commit messages. It supports different change log conventions and formats, depending on how detailed the message should be. For instance, the logs may contain a convention to be in the following format – `<type>(<scope>): <summary>`. These conventions either become more complex or simpler, depending on the choice of the framework developer.

- **commitlint** (`github.com/conventional-changelog/commitlint`): Another tool that helps validate the commit message structure and adheres to a certain commit convention. It is capable of validating the messages as strings through an interactive CLI or programmatic usage.

- **Changesets** (`github.com/changesets/changesets`): A more involved solution to generate change sets for simple and multi-package code bases. This helps when a framework is spread across many sources.

After you look into using these tools for your framework projects, you will find that there are common themes to all of them. Most of the time, these tools have different approaches to configuration and structures. However, with so many options to choose from, using any of these can help you save time by providing important information about the progress of your framework to your users. For instance, the Ember.js project has an easy-to-follow change log (`github.com/emberjs/ember.js/blob/main/CHANGELOG.md`) that helps developers keep up with updates. If you explore some of the change logs, including the Ember.js one, you will see different types of versioning that are maintained throughout the development cycle, which is what we will cover in the following section.

Versioning

Maintaining proper versioning for your framework can be a challenge. Keeping proper versioning schemes helps stakeholders understand the compatibility of their code with new iterations of the framework. If a project is properly versioned and is strict in defining its versions, then downstream consumers can be confident when they update the framework dependency. Any changes in the upstream framework versions can cause breakages in the existing components and cause havoc in the already-built applications. There are many cases, especially in older software, where framework users are asked to be cautious updating to the latest versions of a framework, even if projects are known to have security vulnerabilities. Some examples include drastic changes from AngularJS to modern versions of Angular or major version changes of JavaScript tooling, such as **webpack**. In addition, if a project frequently breaks a versioning contract, then the users of the framework will slow down their upgrade cycles and potentially avoid using particular projects in the future.

Luckily, the software community has created standards around versioning, such as **SemVer** (`semver.org`) and **Calendar Versioning (CalVer)** (`calver.org`), which can help define proper versioning contracts. These standards can help define the framework release process and should be documented somewhere in the framework documentation.

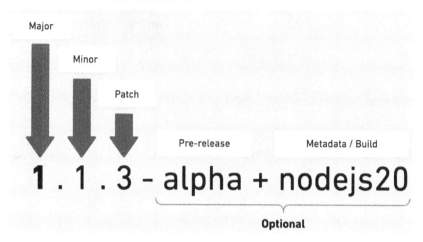

Figure 9.2: Semantic versioning

Semantic versioning, as shown in *Figure 9.2*, is the most common system adopted across many projects nowadays. It consists of three required and one optional components:

- First, we have the **Major** version, which signifies changes that are incompatible if a user updates from an earlier major version

- The **Minor** version means new features are available, and the release is still backward-compatible with earlier versions

- The last required component is the **Patch** version, which means that there a bug fixes that are also backward-compatible with earlier versions

- The **Optional** component toward the end of the version can consist of pre-release names, with the addition of metadata or build numbers

The `semantic-release` (`npmjs.com/package/semantic-release`) package uses SemVer and attempts to simplify versioning for JavaScript projects by utilizing the code base commit history, determining whether a version of the project should be a *major*, *minor*, or a *patch* release. It works as a utility during an automated release step.

The alternative CalVer format defines its structure using dates in the `YYYY.MM.DD` format. It can be useful if project releases are based on the calendar year. Ultimately, the framework maintainer decides the set version, and it is a big responsibility to correctly set these versions and ensure that downstream users do not have their application builds broken.

Simplifying releases

So far, we have seen many tools that automate different parts of the release process. We focused on versioning, feature feedback, and change logs. We can introduce tools to simplify our framework's new release workflow further. This type of tooling aims to ensure that all the release process tasks are successfully executed and requires releases to be consistent across the board.

```
Publish a new version of componium-test (current: 1.0.3)

Commits:
- Adjust fake interface  197e638
- Update changelog  35d603d
- Add restore to mocking library  30b3be5
- Add mocking library  38a6b4c

Commit Range:
1.0.3...main

Registry:
https://registry.npmjs.org/

? Select semver increment or specify new version (Use arrow keys)
> patch          1.0.4
  minor          1.1.0
  major          2.0.0
  prepatch       1.0.4-0
  preminor       1.1.0-0
  premajor       2.0.0-0
  prerelease     1.0.4-0
  ─────────────
  Other (specify)
```

Figure 9.3: Publishing a new release

Tools such as **release-it** (github.com/release-it/release-it) and **np** (github.com/sindresorhus/np) make it possible to ensure that the release process goes smoothly. *Figure 9.3* shows the np tool in action when publishing a new version of the Componium test. These tools ensure that the following tasks are completed for your project:

- Any prerequisite scripts are executed, which could include formatting and linting files.

- The required release tests run and pass. If the tests fail, then the release process is aborted.

- Bumping the version number using the maintainer choice or based on some other criteria, such as using the *semantic-release* package that we saw earlier.

- Publishing code to a specific registry. For internal projects, this could be an internal source; for public projects, this means uploading the source code to a public registry. This will probably be the source from which the framework users get the latest code.

- Pushing the necessary tags to the code repository with the same version number as the one published to the registry.

These are just some of the steps that are generally run, but for more complex projects, many of the steps can be tweaked to accommodate the needs of the code base. Usage of these tools depends on how well they fit into your workflows. It's a good idea to start simple and find tools that work right outside of the box. As the framework project grows, you will find yourself mixing and matching different tools to craft your own approach to the release process. For example, *Chapter 3* mentioned the *ng-dev* tool used for internal Angular development. In the internals of that tool, the team utilizes the *np* command-line tool for the release process. The `release-it` package offers some extra features, suitable for projects that live inside a mono-repo code base or require further configuration.

Maintenance tool showcase

The book code repository in the `chapter9` includes a collection of tools that you can quickly try out and see how effective they are. Your framework projects can integrate the included or similar tools, thus improving the framework development workflow. The `chapter9/commitizen` directory consists of a project that uses the Commitizen package to enforce Git commit guidelines for your projects.

The `maintenance-tools` directory showcases several Node.js utilities used for framework maintenance. To see the available scripts, make sure to run `npm install` and then `npm run dev`.

In the next section, we will reach the final milestone of the release process, which is combining the tools we have seen so far and the continuous integration environment, making it possible to publish a new release of a framework with a single click of a button.

Continuous delivery

Configuring and maintaining the infrastructure for framework release and other tasks involves following best practices and adapting tools from the DevOps methodology space. This involves mixing software development tasks and IT operations to improve software releases, making them much more easily manageable. Often, integrating with DevOps systems will require learning new technical skills outside of the core technologies used in the framework that you are working on. This includes learning about the latest approaches to automation, secure release processes, and application deployment in DevOps environments.

These days, it is ubiquitous and effortless to set up the **Continuous Integration (CI)** step for software projects, and it is important to do so for framework projects. It is important to use the CI environment to make sure that the required framework tests run in an isolated environment. The CI steps also ensure code quality and help create a good framework development workflow. The **Continuous Delivery (CD)** pipeline is designed to deliver the framework product. It is configured alongside the CI steps to prepare the code changes to be inspected, built, and tested. These pipelines are also configured in both open source and internal environments.

The **Delivery** part ensures that maintainers can prepare new releases of a project, which includes executing the set of tools that were part of the simplified release routines. In the delivery phases, the internal development scripts can run all the relevant tasks to the release, which could include generating project documentation and publishing other artifacts. This release environment also has access to the required credentials to publish code to the relevant registries. During the delivery stages, maintainers can configure all the different types of tools that we have seen in this chapter to automate the process of publishing a new version of the framework software that is produced.

An example of this would be a workflow set up with *GitHub Actions* to release new versions of Next.js. These workflows can be seen executing at `github.com/vercel/next.js/actions/workflows/ trigger_release.yml`, triggered by the maintainers of the framework. Configuring these automated workflows will give you a massive boost in productivity, as it will avoid a lot of manual tasks. This will also give you more confidence in your product because these workflows enforce the high-quality bar for all framework maintainers.

Continuous integration sample

Similar to the information shared about the maintenance tool in the *Simplifying releases* section, you can find a sample of CI configuration in the `chapter9/ci` directory. This configuration can be used for your own projects with the GitHub Actions (`github.com/features/ actions`) and Circle CI (`circleci.com`) infrastructure.

To test out these configurations, you can copy the files from the chapter into your own repository, where you have full access to edit the source of the project.

Another relevant aspect of releasing is licensing, wherein framework authors need to explain the terms of use and distribution for their creations. This is what we will discuss in the next section.

Licensing

As your framework develops, you will find you need to set an appropriate license for the code base. This release procedure can apply to both internally developed projects and open source initiatives. In both cases, you can choose from many different types of license types. For private business-related projects, you can choose a **proprietary license**, restricting the use of the framework outside of the control of the company. This type grants exclusive ownership of the code base and internal projects. A **commercial license** can be useful if you want to sell or restrict the redistribution to only the users who paid for the use of your code. For example, you can find some JavaScript frameworks distributing different editions of their frameworks, such as **Sencha Ext JS** (`store.sencha.com`), which includes community and enterprise versions. The extended enterprise versions can include more support, custom features, and dedicated developer attention.

For open source use cases, there are licensing options as well. The website at `choosealicense.com` supports software developers and helps them figure out the needs behind open sourcing their work. You will find many popular open source projects using the following licenses:

- **MIT**: This is a very permissive license that allows commercial use, distribution, modification, and private use. It's based on the conditions that you retain copyright notices and avoid any liability or warranty from your code.

- **The GNU General Public License (GPL)**: This is a copyleft license that offers similar commercial permissions to the MIT license but is more detailed around patent and distribution rules. However, with this license, there are conditions for disclosing the source code.

- **Apache License 2.0**: This is a permissive license similar to MIT but with additional limitations on trademarks and providing patent use cases. If anyone changes the code of your framework under the Apache license, then they need to declare the changes.

As a framework author, it is important to decide on the license that all contributors will adhere to while contributing to a project. The process of changing the licenses down the road takes quite a bit of effort because all the contributors have to re-license their code under the new license. It is also important to remember the license types of libraries you use within your framework.

This concludes the section on release process items, which included gathering feedback, notifying users of new releases, and helping optimize those releases. Now, we are ready to move on to additional topics of maintenance that focus on the long term.

Long-term maintenance

So far in this chapter, we have looked at maintenance tasks that occur as a framework progresses through the development life cycle, covering topics around the initial or following feature updates. However, there are additional unique aspects of framework development that are part of the longer-term upkeep. To focus on a few, we will explore the topics of security, dependency management, evolving feature compatibility, and more.

Security

The approach to web application security has changed in recent years. More tools and solutions are now available in the security space that try to protect the whole development workflow. When users choose a framework to suit their needs, they also have certain security expectations from it, especially if the framework is built for purposes that handle critical data and user input. As you maintain your framework, you can expect to receive security bugs and patches that address security vulnerabilities. The bug bounty programs websites, such as **HackerOne** (`hackerone.com`) and **Huntr** (`huntr.dev`), focus on protecting software and can reach out with vulnerability reports filed against your framework. Both internal and open source frameworks can receive reports, and as a maintainer, the expectation is to fix the known vulnerabilities to maintain a strong security posture.

The vulnerabilities that are created could be assigned a **Common Vulnerabilities and Exposures** (**CVE**) identifier. For example, look at Electron's *CVE-2022-29247* (`nvd.nist.gov/vuln/detail/CVE-2022-29247`), which reports a vulnerability in the framework's process communication. It outlines the fixed versions of the framework and the risk score.

To stay on the offensive and reduce the risk of vulnerability, you can follow the following strategies:

- **Document dangerous APIs**: Invest time in writing up documentation to highlight potential APIs that can present danger when misused. In a server framework, this can involve explaining how to protect against dangerous request payloads. In the frontend, issues can result from unsafely rendering HTML or failure to sanitize URLs or other types of input. For instance, the Vue.js project has a best practices guide that includes information on this topic: `vuejs.org/guide/best-practices/security.html#what-vue-does-to-protect-you`. This strategy also applies to non-application frameworks as well.

- **Security audits**: These types of audits can help run a framework against common attack vectors or specific vulnerabilities that can affect the framework feature set. During this process, your code can be audited by an internal security team or a third party to find potential issues. The goal is to find attack vectors that could make your framework cause harm even when used properly. For application-level frameworks, there exists the **OWASP Application Security Verification Standard** (**ASVS**), which outlines 70+ pages of technical security controls to ensure secure development. These controls can be found in several languages at `owasp.org/www-project-application-security-verification-standard`.

- **Update dependencies**: Relying on external modules and libraries introduces security risks when vulnerabilities get discovered in the underlying code. From what we have seen in Componium and other JavaScript frameworks, there are a lot of external dependencies that projects count on. Recently, more and more security scanners, such as Socket (`socket.dev`) and Dependabot (`github.com/features/security`), have become available to track down JavaScript vulnerabilities in particular and inform the maintainers to fix them. However, these scanners cannot fix the issues and create releases, so it is still up to framework developers to keep up with all the dependency updates.

- **Define a security policy**: Creating a security policy can outline how security issues are reported against your project and provide guidelines to contact maintainers to fix particular issues. Many frameworks from *Chapter 1* define a `SECURITY.md` file, where the security policy is documented. It can usually be found in the root directory of the project; the one for the `express.js` project is at `github.com/expressjs/express/blob/master/Security.md`.

There is always a lot to keep track of in terms of framework security maintenance, but even investing a bit of time into the security posture can help reduce your burden and improve sufficiently benefit your project. Keeping up with security tasks is also relevant for the dependencies of your project. The next section focuses on managing the dependencies that can affect your project in different ways, including the security aspects of the project.

Dependencies

In the long term, managing the dependencies of a JavaScript framework can be a very involved task. The libraries and modules that are relied upon can become outdated or unmaintained, and this goes beyond being affected by security issues. As an ecosystem moves forward, framework developers need to keep an eye on some of the stale modules that are used internally. The lack of updates to dependencies can be limiting when a lingering bug fix depends on a component outside of the framework code base. If the dependent package is fully abandoned, creating your own copy and attempting to fix the issue is a good idea. The other option is to migrate to a similar package or rewrite it independently. Dependencies can also break compatibility in some way, and it will be a maintenance task for the framework author to refactor the usage of that module to restore compatibility.

A more positive turn of events can be additional features being added to the libraries that are used within a framework project. In such cases, the project and its users can benefit from the improvements. These enhancements can come with the addition of new, exciting features or potential performance optimizations.

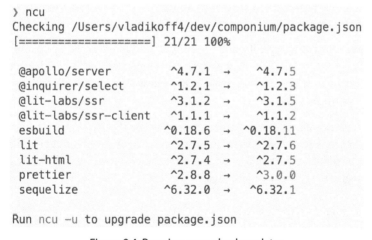

```
> ncu
Checking /Users/vladikoff4/dev/componium/package.json
[====================] 21/21 100%

@apollo/server        ^4.7.1   →    ^4.7.5
@inquirer/select      ^1.2.1   →    ^1.2.3
@lit-labs/ssr         ^3.1.2   →    ^3.1.5
@lit-labs/ssr-client  ^1.1.1   →    ^1.1.2
esbuild               ^0.18.6  →   ^0.18.11
lit                   ^2.7.5   →    ^2.7.6
lit-html              ^2.7.4   →    ^2.7.5
prettier              ^2.8.8   →    ^3.0.0
sequelize             ^6.32.0  →    ^6.32.1

Run ncu -u to upgrade package.json
```

Figure 9.4: Running npm-check-updates

We can count on some dependency management tooling to keep track of dependencies for us. For example, *Figure 9.4* shows the **npm-check-updates** (npmjs.com/package/npm-check-updates) tool running via the ncu command line, which tracks down some of the dependencies that have newer versions. Alternatively, automated CI tools can also produce a similar report.

Due to the nature of the ecosystem, keeping track of dependencies in JavaScript projects is especially hard, so having a certain strategy, either through using tools or minimizing the number of dependencies, can be quite useful for a framework project of any size.

Dependency management fits into the larger theme of feature coverage. As a project progresses, framework designers alter how features are structured and remove the unused ones. This is something every maintainer needs to consider for the long-term strategy of their framework project.

Feature coverage and deprecation

Earlier, in the *Development cycle* part of the chapter, we used the SDLC and defined processes to drive feature development. With the long-term outlook of a framework project, it is useful to keep a good coverage of features that are practical and features that are not used by any stakeholders. When we consider user feedback, we also need to make sure that a feature is worth adding from a long-term maintenance point of view. This is where it is worth considering the quick wins of adding more features versus the project's long-term vision. In a similar fashion, deprecating features of a framework might mean cleaning up the less relevant components. Usually, this would involve a lengthy process of creating a migration path for existing users and providing an alternative solution. Otherwise, the project would lose some of its credibility over time. To avoid expanding the hurdles of complex feature management, frameworks create extension interfaces that allow for feature expansion without bloating up the core functionality. We have seen examples of this in several projects. For instance, the Componium server allows custom middleware functions to intercept the requests based on the `express.js` behavior. Vue.js is a frontend framework example that offers a plugin interface for functionality, which cannot be bundled in the core framework: `vuejs.org/guide/reusability/plugins.html`.

Performance optimizations are a type of feature optimization that often end up spanning over a long term. Frameworks may find bottlenecks or slowdowns through user feedback or particular use cases. This is where performance initiatives, which span many releases and large refactorings, can be useful to develop a more optimized product.

In this section, we covered some of the long-term issues and tasks that may come up during the lifespan of frameworks. Other maintenance undertakings that we did not cover could be solved by familiar patterns, including introducing particular tooling, utilizing an external service, or relying on existing software methodologies to reduce the maintenance burden.

Summary

In this chapter on framework maintenance, we learned and reviewed some new and familiar topics – the development cycle, the release process, and maintenance tasks. These three topics enable us to successfully maintain a JavaScript software project over long periods of time. Part of the reason we dived into the details of these subjects is to enable you to create your own maintenance workflow, with your choice of tools and techniques.

When we looked at the steps of the development cycle, we scoped it down to the specificities of JavaScript framework development. Alongside that topic, we learned about the RFC process and found ways to get valuable feedback from the users of our frameworks. Furthermore, we focused on the release process, which included learning about structuring our approach to versioning, licensing, documentation, and so on. Finally, the long-term maintenance tasks included preparing for events that occurred previously in other JavaScript projects. These included topics such as dependency management, dealing with security incidents, and handling out-of-date features.

Overall, we have captured the essence of framework maintenance, and this should give you a good foundation to explore the other aspects of maintenance that are present in other projects. I encourage you to examine other frameworks. For instance, by looking at the open source frameworks from *Chapter 1*, you can find other examples of tools and techniques used in the maintenance of those projects.

In the next and final chapter, we shall circle back to all the key fundamentals from this book and conclude our JavaScript development adventure by discussing the best practices of this topic.

10
Best Practices

Within this final chapter of the book, it is time to wrap up our journey, with several vital topics around general JavaScript framework development, and take a glimpse into the future of frameworks in this ecosystem. Over the preceding chapters, we have dissected real-world examples and built a robust knowledge base centered on project maintenance and organization. Harnessing that practical knowledge, in this chapter, we will focus on the current status quo of frameworks and examine several predictions to see where innovation in this space is heading in the future. The overall goal is to understand the best practices of JavaScript framework development that are prevalent today, as well as to explore some of the future patterns as they emerge.

The essential takeaways from this chapter will be around bridging the gap between where we are today in the framework development space and what kind of solutions framework authors will build in the near to long-term future. This in-depth exploration will cover the following topics:

- **The common themes of frameworks**: The first part discusses several architectural patterns and common choices across many of the framework projects we have seen in this book. Elements such as modularity, approaches to code bases, standardization of best practices, and performance-based design are cornerstones of effective and robust frameworks. With these elements, we will be better equipped to anticipate innovation in this sphere of software development.

- **Future of frameworks**: We will see what factors will affect how frameworks evolve as time passes, focusing on themes dealing with developer experience, addressing full-stack complexity, and highlighting potential new approaches to development. This section highlights the potential paradigm shifts in development approaches poised to redefine the industry's trajectory. Considering and researching new trends and techniques is important as you start building new software for the public.

- **Additional development considerations**: Finally, to wrap up our adventure, the section on additional considerations turns the spotlight on important factors such as time investment, financial backing, and overall software support. These factors, often overlooked as projects progress, significantly shape the process and outcome of framework projects. These additional considerations are important for any type of framework developer.

Following these topics on *best practices*, this chapter highlights the enduring principles of JavaScript framework development – taking common themes into account, looking into the factors that will shape the future of similar projects, and looking at the auxiliary considerations that must be taken into account. Understanding these *best practices* is the key to unlocking your potential to comprehend and influence the development trajectory of your own JavaScript frameworks. These skills will remain future-proof, regardless of the changing tech landscape, as you further dive into the development of your project. The first section on common themes in development looks at several examples of concepts reused in framework development today. Let's begin.

Technical requirements

Similar to *Chapter 9*, the `chapter10` directory consists of several sample projects that showcase tools related to framework best practices. Refer to the `chapter10/README.md` files for a guide that documents the contents of the child directories of those chapters. The technical requirements are similar to the other chapters, such as the use of Node.js 20 or higher.

Common themes

Looking at the current state of framework projects in the JavaScript ecosystem, we can see stability, vibrancy, and chaos in the JavaScript framework field. For instance, we have seen many projects utilizing an approach to build on existing primitives, such as many of the frameworks using the React component library as the foundation for component architecture and rendering in the browser. At the other end of the spectrum, projects are created from the ground up, reinventing the approach to rendering in the browser or solving particular challenges of software development with JavaScript. This section explores similar common themes that occur across many projects. Knowing these particular commonalities helps framework developers stay in touch with the rest of the ecosystem and develop more cohesive projects.

When we take a zoomed-out view of the current state of all these projects, at one end of the spectrum, we find large, established frameworks that underpin numerous high-traffic applications and sophisticated tooling. With every release, these frameworks augment their maturity and stability. For instance, Electron sits on the throne as the most popular framework for utilizing web technologies for application development. Each new version steadily improves its design and addresses its performance metrics. Conversely, a set of evolving projects generates fresh, new ideas in the JavaScript community. These newcomers, whether introduced as public resources or crafted for internal business needs, inject a dose of novelty and versatility into the ecosystem. For example, *Svelte* and *SvelteKit* challenge some of the established paradigms and sway certain developer mindsets toward a different approach. As the approach to web application architecture changes and evolves, this whole spectrum entices excitement, opportunities, and new technological advancements for all JavaScript developers.

In the *Framework showcase* section in *Chapter 1*, we charted the evolution of framework development across an expansive timeline. From their beginnings as specialized solutions for specific tasks, such as their initial focus on single-page apps, frameworks have evolved to become all-encompassing development platforms. A modern-day framework's functionality is exponentially broader and richer, including solutions for full stack needs and beyond. In the practical example of the development of the *Componium* framework in *Chapters 6, 7*, and *8*, we saw a reliance on the diverse set of packages and abstractions to construct this comprehensive framework to form a fully fledged full stack system.

Modularity

Modularity is one of the common themes we can devise from many of the projects mentioned in this book. The modularity concepts apply to JavaScript projects in many ways, and they are also specific to JavaScript projects, compared to projects in other programming ecosystems. The modular approach to development has been fueled by the structure of web applications and the package structure in registries such as npm, backed by the package.json format. In comparison, when looking at programming languages such as Python, they depend on external project dependencies but lack a standardized method to effectively handle these dependencies. JavaScript is in a unique situation, where the frameworks use many internal and external modules. This approach is both beneficial to development velocity and burdensome from a maintenance point of view.

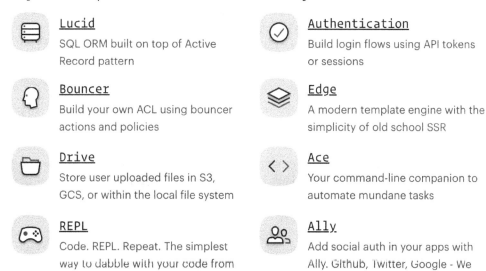

Lucid
SQL ORM built on top of Active Record pattern

Authentication
Build login flows using API tokens or sessions

Bouncer
Build your own ACL using bouncer actions and policies

Edge
A modern template engine with the simplicity of old school SSR

Drive
Store user uploaded files in S3, GCS, or within the local file system

Ace
Your command-line companion to automate mundane tasks

REPL
Code. REPL. Repeat. The simplest way to dabble with your code from the command-line

Ally
Add social auth in your apps with Ally. Github, Twitter, Google - We have drivers for all

Figure 10.1: The modularity of AdonisJS

Figure 10.1 depicts an example of modularity in *AdonisJS*. The figure presents the outline structure of the first-party packages of that framework. The AdonisJS framework packages are well-organized and decoupled for the sake of developers, who can choose what fits their use cases better. Most of the packages are installed from the Adonis namespace – npm install @adonisjs/ally. They are later configured using the framework's command line tool, called **Ace**. In the AdonisJS core, the code base also relies on several modules for development and end user features: `github.com/adonisjs/core/blob/develop/package.json`. This is just another example of a common theme we see across all of the projects in the ecosystem; this approach will likely not change any time soon as new projects become more popular in the community. As a framework developer, you would embrace the available existing code and structure your projects to consist of modules as well.

Testing out the modularity

The `chapter10/adonisjs` directory consists of a sample AdonisJS project. In the project directory, you can run `npm install` and then `npm run dev`.

Once the project is running, you can open the following address in your browser: `http://127.0.0.1:3333/`. With the example app running, you can install and use additional modules, such as the one listed earlier in this section, via `npm install @adonisjs/ally`. You can find more modules and packages at `adonisjs.com`.

The increased complexity of architectures contrasts with the benefits of modularity. For instance, the more isolated the modules that a framework consumes or exposes are, the more you have to think about coupling and how all those modules fit together. More recent challenges deal with keeping dependencies up to date or tagging new releases of a framework's separated modules. Framework developers have to invest a lot more time in managing the dependencies they consume and produce, which is not something that is going to change in JavaScript platforms any time soon.

Evolving design

Another common theme among many JavaScript frameworks is the concept of evolving design. Similar to systems in other languages, the frameworks that are built in JavaScript need to evolve over time in response to the changing environment. The changing factors could include technological changes or advancement, new industry trends, or catching up with competing frameworks. In the case of JavaScript, these factors are advancements to web browsers, Node.js APIs, runtime improvements, and so on. Well-defined abstractions and thoughtful architecture can help you to adapt to these changes, without having to do drastic refactoring in framework projects. Several examples come to mind that show evidence of such drastic changes in environments in which JavaScript frameworks exist, such as the introduction of web components and related modern APIs for frontend frameworks. Some of the frameworks chose to embrace the new standards or integrate with them to be more compatible with the evolving web.

Another example is migration to the **ECMAScript module** (**ESM**) format. First, projects had to adapt to a third-party module system, such as CommonJS or AMD, or implement their own system. Then, as the official JavaScript module definition was created, it was up to the projects to embrace the new ways of structuring code bases. With all the benefits of ESM, such as the static module structure and improved syntax across all JavaScript environments, there were still intricate incompatibilities across some of the use cases. It is up to framework authors to determine and evaluate the ESM support for their projects. For example, since version 16, the Angular project started off supporting ESM modules as a developer preview and offering it as an option. This enabled the project to expand the feature set with dynamic import expressions and lazy module loading. In addition, the change improved the build-time performance for application builds. It also allowed the framework to use more modern tools such as *esbuild*, which was also used in the Componium framework.

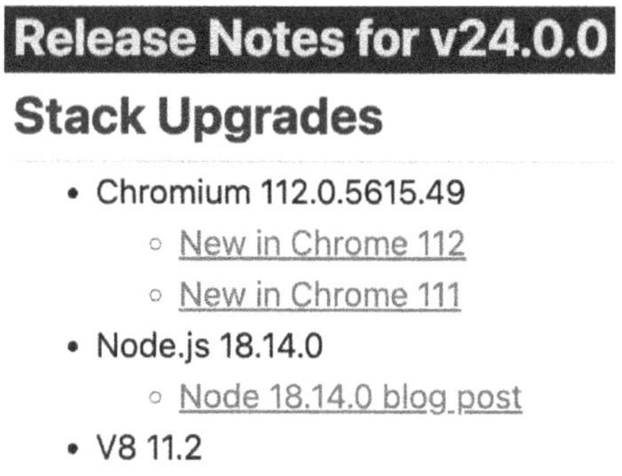

Release Notes for v24.0.0

Stack Upgrades

- Chromium 112.0.5615.49
  - New in Chrome 112
  - New in Chrome 111
- Node.js 18.14.0
  - Node 18.14.0 blog post
- V8 11.2

Figure 10.2: Partial release notes for Electron.js

For an application framework such as Electron, the framework feature offers changes on every major release. This is because the major releases often track changes to the underlying Chromium, Node.js, and V8 changes, as seen in the release notes in *Figure 10.2*. With new fixes and features in those components, what the framework offers evolves as new releases get shipped to the public. This is a compelling example of how a project leverages continuously evolving improvements from external dependencies.

New features such as server-side rendering have been added to many frameworks as new ways of hydrating and producing views on the server side became available. Existing projects add new capabilities to allow features such as server-side rendering to fit into the existing API surface and architecture. Across the board, we can see these similar changing trends in projects of all kinds. Thoughtfully adapting to the latest trends avoids project stagnation and allows us to keep up with the latest JavaScript advancements.

Minimalistic approach

Another common theme is the minimalistic approach to framework design. Some frameworks may choose to focus on simplicity and minimal architectural churn. In these instances, the number of dependencies and complexity are greatly reduced. The more simplistic frameworks can be effective in low-resource environments and for projects that do not require large framework overhead. In JavaScript projects, these frameworks usually aim to be a simple API with a tiny file size footprint and primarily focus on a specific isolated feature set. If the feature set fits the requirements, choosing this type of approach can reduce the amount of resources for framework development and produce a much cleaner, simpler interface for stakeholders.

To highlight some of the examples, from the frontend perspective, *Preact* (`preactjs.com`), as a library, takes a minimalistic approach, offering a 3 kilobyte alternative to the React library. It can be used in a minimalistic framework for frontend rendering. A backend example would be the *Hapi.js* project from *Chapter 1*. It focuses on the features of building API endpoints really well; if you look at the source of the framework, you will find fewer than 20 files within its core.

It is important to remember that you can build minimalistic frameworks for your needs and don't always need complex tooling and a large feature set. This type of approach is not always about the final framework size or the number of files in a project, and it can also be used as the guiding principle as you make decisions during framework development. In many senses, when these types of projects are used for real-world projects, they are just as capable of achieving great results.

Build tooling

In *Chapter 3*, the *Framework compilers and bundlers* section showed examples of build tools. These are commonly shipped together with a framework to enable optimized application bundle outputs. Some frameworks also utilize different build tooling types or have flexible options to allow stakeholders to choose their build tools. Nowadays, the trend is for build tools that make it easier to produce outputs for many JavaScript environments, with a focus on speed. The other principal aspect around the combination of frameworks and build tools is the cooperation of many developers on improving build tooling workflows, enabling more use cases that fit different project requirements.

The additional benefit of build tools found in popular tools such as **webpack** is the enforcement of sound patterns and warning developers when application output is not suitable for a client or server environment. For instance, packaging tools can warn developers when a packaged bundle is too big for browsers to load or when it might not conform to the client environment that it targets.

The build tooling drives constant refinements in frameworks, through optimizations and advancements, and contributes to performance improvements for JavaScript applications, which is the next common theme that we will highlight.

Performance improvements

The emphasis on continuous performance improvements and benchmarking is another theme that is quite typical among JavaScript frameworks. Depending on the framework environment, the optimizations focus on consuming fewer computer resources, improving load or response times, promoting slicker user interaction, expanding rendering capabilities, and so on. These types of optimizations require greater knowledge of the JavaScript language and the ability to optimize the existing code. The process of optimizations also gets more and more complex for larger projects.

In the many years since the early JavaScript frameworks, there have been quite a few established benchmarks and benchmarking tools. However, in many cases, these types of benchmarks cannot truly benchmark real user behaviors or real-world application use cases. Just like other types of software benchmarking, these performance testing systems establish a standard test to compare implementations in different scenarios. Even though, in many cases, the results can be flawed, they can offer some insight for stakeholders trying to find a framework useful for their use cases. In addition, even after many years of benchmarking battles, frameworks still want to showcase that they are ahead in certain feature sets compared to the competition.

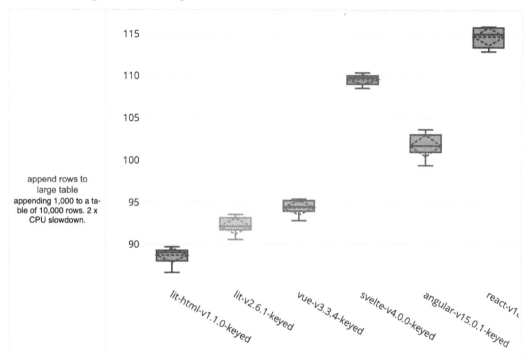

Figure 10.3: An open source js-framework-benchmark

Figure 10.3 previews a filtered example of results from the **js-framework-benchmark**, which is a benchmark focused on common list operations. It can be found at `krausest.github.io/ js-framework-benchmark`. This particular benchmark focuses on frontend solutions. If you develop a frontend framework, you can add your code base to the list of existing frameworks to see how it compares. These types of tests can showcase how the framework will work for applications that include complex lists of data, including updates to the rows of data. This benchmark also dives deep to take into account memory usage, load time, and framework size.

Other types of frameworks require different types of benchmarks. For example, there is a comprehensive comparison at `techempower.com/benchmarks` for backend web frameworks. These tests outline a particular load test for a server-side framework and include projects from other programming languages. These types of frameworks focus more on the latency of requests, throughput, interaction with database queries, and so on. Aspects of the framework, such as the throughput of requests, demonstrate how a server running a particular framework will act under heavy load.

Different types of frameworks can utilize similar tests that exist in their framework space. If there are no existing types of tests for your kind of framework, then you can establish your own benchmarks and keep track of the results from release to release, focusing on improving the numbers that are relevant to the performance of the most relevant feature sets.

The common themes of today's frameworks can vastly influence your framework project, but to create innovation in the systems you build, there is more to learn from upcoming trends. In the following section, we will take a look at future types of trends that influence new JavaScript projects.

Future of frameworks

The JavaScript ecosystem is still one of the most vibrant among other programming languages, and it is bound to grow and expand into new areas. As this happens, we will start seeing future innovations that will affect how frameworks are built and how they are utilized. The patterns used by the developers we see today will also change as time progresses. Keeping up with the industry tendencies and looking at what is ahead will help you build better systems and incorporate the latest trends into what you build. In this part, we take a look at some potential areas that framework evolution and improvements can potentially head toward.

Developer experience

We have seen how important the **developer experience** (**DX**) can be in a framework. In the future, the DX factor will even further differentiate good frameworks from great ones. Providing additional tooling and finding ways to reduce complexity makes it much easier to start building with a framework. In recent years, we have seen tactics to enable an end-to-end framework DX, helping stakeholders with each step of the process of building applications. In company-backed projects, such as Next.js, those who use the framework experience a fully encompassed workflow from the beginning of development, all the way up to the deployment of applications. The concepts of amazing DX fits really well with the concepts of reducing the overall complexity of using a framework. Here are a few themes that will be the focus of DX improvements in the future:

- **Reducing the learning curve**: Framework authors will continue investing in making interfaces more approachable, especially for new developers. In both server and frontend environments, this could mean further aligning with the structure of web APIs, which helps avoid introducing new types of abstractions that have unique interfaces. A tip to a future framework developer would be to contemplate how a learning curve could be simplified for your projects.

- **Simplified configuration**: Frameworks will take further steps to simplify out-of-the-box configuration, and picking options that are the best fit will help with development. These will include more sensible defaults, which we saw in *Chapter 7*, where certain parts of the framework focus on a simpler configuration. For your own projects, you should take care of every configuration option you introduce and avoid overwhelming users with the necessary configuration steps.

- **Improved testing**: Projects will continue to make writing and executing tests easier, further focusing on simplifying the testing of hard-to-test components. For example, as new end-to-end testing frameworks have been developed, these projects have addressed the common developer annoyances of flakiness in tests and lack of debugging tools in CI environments. An emphasis on simplifying testing in your frameworks can help the day-to-day experience of your users, as writing tests is a big time sink.

- **More built-in tooling**: In the past, many types of frameworks have relied on external tooling, usually by including those tools in developer dependencies. An example of this can be seen in the Angular `package.json` source code. Future projects will likely abstract away some of the tools they use in the background, allowing for more control over the DX. Furthermore, projects will also build more of their own tooling, using the existing primitives available in the ecosystem.

- **Focus on performance**: Similar to the performance improvements described in the *Common themes* section, projects will continue pushing the limits of the JavaScript language and runtimes to find ways to deliver improvements on aspects such as rendering, latency, and other relevant metrics. With the recent emphasis on performance in the frontend space, developers adhere to indications of quality that can improve the experience of web applications. These metrics and concepts can be found at `web.dev/learn-core-web-vitals`.

- **Additional flexibility**: Frameworks will continue adding more options to support different environments. For instance, backend solutions will expand integrations with more databases. Application frameworks, such as Electron, will take advantage of the latest APIs available to them in the browser runtime and within the operating system. For the frameworks you build, you should aim to strike a balance between flexibility and the number of use cases you support. You should only add additional support for use cases if you have enough resources to maintain this large set of features.

- **Packaging and bundling improvements**: There have been significant changes in how frameworks package their application code. This focus on improved packaging techniques will only continue with a priority on speed and enabling more options to bundle code in different ways. For framework developers like yourself, it is important to keep up with the latest improvements to these types of bundling tools, as they can benefit how you approach the packaging of the system you build.

This list does not cover all the potential directions of what JavaScript systems will head towards in the future. However, following the established trajectory of the existing projects helps to highlight where you, as a framework developer, can deliver significant impact and make a difference for your users.

Embracing new technologies

Besides the developer experience improvements, new technologies are made available in places where JavaScript applications execute. As highlighted in *Chapter 1*, web assembly will play an important role in enabling the next generation of solutions for computation-heavy tasks. This has the potential to expand the code base of the existing frameworks to include lower-level languages. It will also require maintainers to expand their domain knowledge beyond just JavaScript code. On the theme of other languages, the usage of TypeScript will continue to grow, as its benefits are very worthwhile to framework authors. However, consumers of frameworks will still be able to use the *built* versions of frameworks that were originally transpiled from TypeScript.

Additional improvements to browser engines will also drive performance boosts and new capabilities in places where a framework relies on web browser engine behavior. One outlet to get inspiration from is the `blink-dev` mailing list at `groups.google.com/a/chromium.org/g/blink-dev`, which highlights the upcoming changes that will end up in Chromium and Chrome browsers.

Package management is also an area where a lot of new technological changes will occur. Almost every project relies on package management tooling, such as `npm`, to resolve and build its code base. Package managers are so heavily involved in the development workflow that any improvement will impact JavaScript development drastically. The future will bring better versioning, dependency resolution, enhanced security of projects, and so on. The evolution of package managers should also make framework organization and development easier, allowing authors to lay out their projects in more developer-friendly ways.

Another big new instrument is the use of AI-powered models for the benefit of all the framework aspects, explored in the following section.

The use of large language models

In recent years, the advancement of **large language models** (**LLMs**) powered by different neural networks has affected software workflows and tools. As a step in framework development, the use of artificial intelligence tooling will likely become more prevalent. Similar to the importance of including documentation for project releases, developers will bundle a trained model specifically for their framework. This model can be aware of the project's public APIs, internals, and documentation. The intertwined models can be involved at different stages of the development process. For instance, as you refactor code, an AI assistant powered by the trained model can provide you with suggestions for potential changes. As these model integrations become increasingly refined, they could also take over some of the more repetitive or redundant tasks involved in project maintenance.

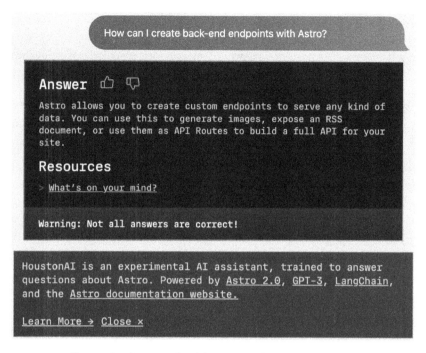

Figure 10.4: An example of the Astro AI framework assistant

Using a real-world example, as shown in *Figure 10.4*, the all-in-one framework **Astro** specifically trained a model to answer questions about a project. This developer tool is made available for public use, solving potential support and integration hurdles by allowing users to ask questions about the project. The tool is available at `houston.astro.build`. This type of tooling can accept a query about certain features of the project, such as creating API routes and endpoints. The response consists of a text response gathered from a mix of documentation and the conversation model. The final resources link directs a user to `docs.astro.build/en/core-concepts/endpoints`, which has information about configuring the endpoints for applications that utilize this framework.

Connecting these similar ideas of using trained AI models for the sake of better developer experience, trained LLMs can be used for scaffolding purposes and within the framework tooling itself. In *Chapter 8*, we have seen examples of generating code within the Componium framework; a potential future enhancement can be prompted for a query from an application developer about the component they are trying to build. The later generated code can save a lot of time for developers, reducing the amount of documentation they have to read and the amount of code they have to write. Creating and understanding these LLM-powered integrations would be an additional step to the full stack developer paradigm. Integrating these types of LLMs at different stages of framework development and maintenance will change where framework authors spend most of their time within a project.

With the current and future themes covered, let's explore what other aspects we should consider that fit into the development of new framework projects.

Additional considerations

In *Chapter 5*, we explored a wide range of framework considerations, specifically affecting the technical architecture. As creating a new framework project, even with a totally unique technical feature set, is a big decision, let's examine a few additional considerations. Outside of the technical challenges, this examination will help us decide whether starting a project from scratch and maintaining it for a long time is worth an investment. Successfully producing a framework project with a lot of stakeholders and targeted users will require both time investment and financial backing.

Besides the technical hurdles and architectural challenges, framework developers need to consider the probable return on investment when embarking on a new project. Looking at some of the established projects in the JavaScript space, many of them have been multi-year investments, with a huge number of developers contributing paid or open source time to enable the feature set. Let's focus on the particular three pillars of time investment, user support, and financial backing in the following sections.

Time investment

Framework developers need to consider a new project's time investment and lifespan. It's not simply about the development phase but also the life cycle commitment it entails, which could affect multiple teams within an organization or the company's larger goals. Open source developers have similar commitments to consider but with lesser responsibility, especially at the early stages of the project. Unlike typical JavaScript application projects, framework development typically doesn't have a linear timeline or a designated end milestone where a deliverable is considered complete. Instead, the process is iterative, with development cycles, testing, refinement, and updating. These themes link back to the chapters on maintenance and the development life cycle of new features.

A framework project will demand more than a sizable initial coding investment created by one developer or a whole team. The ongoing updates necessary to adapt to ever-evolving requirements and technological conventions will require equally weighty commitments. Considering the tasks we saw in *Chapter 9*, the maintenance time investment encompasses the need for a comprehensive workflow, involving many developers. When choosing to build a new solution, consider how much time you are willing to invest and what size framework you can successfully execute on. Making a considerate decision will determine whether your framework effort will succeed.

User support

As important as time investment is user support, which often determines the success of a project. Besides producing readable and well-organized documentation, a deep understanding of the challenges and needs of users spans beyond the code base. For all types of framework projects, as a developer, you will find yourself or your team acting as support and troubleshooting framework integration aspects. To give a public example, the Vue.js project has a discussion forum where users publish their technical and architectural questions: `github.com/vuejs/core/discussions`. It is up to the team to respond to these queries and maintain a good community posture. In projects that are internal to the company, framework authors often set up a similar platform to interact with consumers. For smaller companies, this support interaction is much more hands-on and direct. However, in all of these scenarios, the support aspects take up important development time.

The best strategy to reduce this type of churn is to constantly work on improving support workflows. Improvements can include organizing frequently asked questions for later use, creating guides for complicated framework integrations, and making the framework knowledge base as discoverable as possible. Just like the cyclical development process seen in *Chapter 9*, as a maintainer, you will find yourself in an ongoing commitment to support new and older users of your project. For a project to be successful, you have to put your users first and be able to address their needs in an acceptable manner.

All this time investment in additional aspects of framework building requires a monetary budget, which we will highlight in the following section.

Financial backing

Among the complex technical challenges, time investment, and logistics of supporting a project, there is another factor to consider for a framework – that is, the presence of financial backing. Funding a new project will require enduring the ongoing expenses for feature development and infrastructure costs. Additional costs could arise, depending on the environment of your project – for example, if you are trying to promote an enterprise framework to potential customers. In a company setting, where the framework powers the foundational services and products indirectly, the funding can trickle down from the profits of the products.

Most open source projects are often not fully funded but could be sponsored by a larger organization with a big enough budget for projects, which also indirectly help development efforts or advance the agenda of the organization. For instance, the Cypress open source testing framework has a paid Cypress Cloud service that enables a dedicated environment. Conversely, some of the open source developers utilize platforms such as Patreon or GitHub Sponsors to fund projects, getting direct financial contributions from companies and individual users.

These additional considerations can play an important role in planning and executing your project. Depending on your development goals, some of these factors can make or break the framework project you are trying to establish.

Summary

This chapter has brought to attention the three final topics that can help framework developers better understand the efforts of creating new frameworks, under the common themes of JavaScript development. A look at the future of projects also helps developers understand and prepare for what is to come in this particular technological space. By exploring the common themes, you gain knowledge of how modular projects can be and how their design evolves as time goes on, making it easier to make decisions about self-created projects. Similarly, learning about approaches to performance optimizations and unique architectural patterns can be beneficial to prospective framework authors.

In the second section, the chapter highlighted upcoming trends and looked at the future potential of frameworks, offering us a glimpse of what is to come in this space. The next generation of framework authors and maintainers gets to experience the new challenges of the JavaScript environment and refactor existing solutions to make them more capable in new ways. The expectations of a better developer experience and the still unknown possibilities of LLMs keep the framework space super exciting for both experienced and upcoming JavaScript developers.

Toward the end, the chapter highlighted additional considerations that can help developers make the right decisions in terms of starting new projects. These non-technical considerations are much less exciting to the developers, but they are crucial to the framework creation process. This final chapter aims to provide further insights to bolster your understanding of framework development and prepare you for success in your real-world projects.

In conclusion, the goal of this and all the chapters before it was to demystify the framework development methods and make the process of building your own JavaScript frameworks much more approachable to all. With this acquired knowledge, you now have the skills to make better decisions and drive more complex framework-related projects to success. We began this book with our journey, looking at existing projects, then identified crucial parts of framework development, and later concluded by building a full-stack framework from scratch. This includes an exploration of architectural patterns and project considerations.

Let's summarize the topics that we learned about in all these chapters:

- **Knowledge of other projects in the JavaScript space**: The initial chapter showcased how different types of frameworks have emerged and what problems they solve for a larger developer user base. The collection of projects gave us a peek into existing solutions that help solve software challenges.

- **Framework organization and building blocks**: This taught us the basics around abstractions in frameworks and the basics of the building blocks applied to backend, frontend, and other types of initiatives.

- **Architectures and patterns**: This showed us examples of concepts and structures behind the existing projects, such as Angular and Vue.js. This included mentions of additional tooling that helps put frameworks together for consumption.

- **Ensuring framework quality**: This taught us ideas of how frameworks can provide a quality experience through creating documentation, ensuring well-tested components, and how development tooling can help ensure the delivery of good software to its users.

- **Overview of project considerations**: This helped create a plan for development, with aspects to consider before starting work on technical problems.

- **Creating a new testing framework**: This gave us practical experience in creating a practical framework project. This introduced the technical architecture behind a testing system with detailed features.

- **Developing backend components**: This continued the practical approach with a focus on backend development. We outlined an approach to server-side solutions and developer experience of Componium.

- **Crafting frontend components**: This was the final part of the practical approach that specializes in frontend components. It included architectural design, with concepts of reactivity, server-side rendering, and so on.

- **The effort it takes to maintain a project**: This taught us about the tasks that framework developers have to perform daily and cyclically as new features and fixes get added to a project.

- **Learning about common themes of today and the future**: Finally, this part summarized the typical things we see in JavaScript projects today, with additional considerations for new projects and the future.

All this newly gathered knowledge will unlock more possibilities for you in the JavaScript ecosystem, enabling you to be a more effective and acquainted engineer, greatly benefiting your career and the projects you are responsible for. The web application development field is constantly evolving in many ways, which powers this exciting ecosystem of JavaScript frameworks. Even with the availability of flexible building blocks and fully fledged solutions, there is still room for further expansion and improvement – it has never been a better time to develop a JavaScript framework or contribute to one!

Index

Packtpub.com

Subscribe to our online digital library for full access to over 7,000 books and videos, as well as industry leading tools to help you plan your personal development and advance your career. For more information, please visit our website.

Why subscribe?

- Spend less time learning and more time coding with practical eBooks and Videos from over 4,000 industry professionals

- Improve your learning with Skill Plans built especially for you

- Get a free eBook or video every month

- Fully searchable for easy access to vital information

- Copy and paste, print, and bookmark content

Did you know that Packt offers eBook versions of every book published, with PDF and ePub files available? You can upgrade to the eBook version at packtpub.com and as a print book customer, you are entitled to a discount on the eBook copy. Get in touch with us at customercare@packtpub.com for more details.

At www.packtpub.com, you can also read a collection of free technical articles, sign up for a range of free newsletters, and receive exclusive discounts and offers on Packt books and eBooks.

Other Books You May Enjoy

If you enjoyed this book, you may be interested in these other books by Packt:

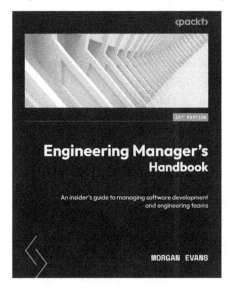

Engineering Manager's Handbook

Morgan Evans

ISBN: 978-1-80323-535-6

- Pitfalls common to new managers and how to avoid them.
- Ways to establish trust and authority.
- Methods and tools for building world-class engineering teams.
- Behaviors to build and maintain a great reputation as a leader.
- Mechanisms to avoid costly missteps that end up requiring re-work.
- Techniques to facilitate better product outcomes.

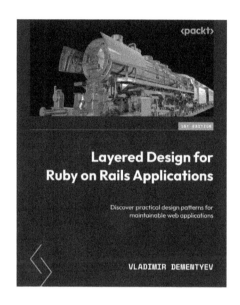

Layered Design for Ruby on Rails Applications

Vladimir Dementyev

ISBN: 978-1-80181-378-5

- Discover Rails' core components and its request/response cycle.
- Understand Rails' convention-over-configuration principle and its impact on development.
- Explore patterns for flexibility, extensibility, and testability in Rails.
- Identify and address Rails' anti-patterns for cleaner code.
- Implement design patterns for handling bloated models and messy views.
- Expand from mailers to multi-channel notification deliveries.

Packt is searching for authors like you

If you're interested in becoming an author for Packt, please visit `authors.packtpub.com` and apply today. We have worked with thousands of developers and tech professionals, just like you, to help them share their insight with the global tech community. You can make a general application, apply for a specific hot topic that we are recruiting an author for, or submit your own idea.

Hi!

I'm Vlad Filippov, the author of Building Your Own JavaScript Framework. I really hope you enjoyed reading this book and found it useful for increasing your productivity and efficiency in understanding and building JavaScript Frameworks.

It would really help me (and other potential readers!) if you could leave a review on Amazon sharing your thoughts on Building Your Own JavaScript Framework here.

Go to the link below or scan the QR code to leave your review:

```
https://packt.link/r/1804617407
```

Your review will help me to understand what's worked well in this book, and what could be improved upon for future editions, so it really is appreciated.

Best Wishes,

Vlad Filippov

Download a free PDF copy of this book

Thanks for purchasing this book!

Do you like to read on the go but are unable to carry your print books everywhere?

Is your eBook purchase not compatible with the device of your choice?

Don't worry, now with every Packt book you get a DRM-free PDF version of that book at no cost.

Read anywhere, any place, on any device. Search, copy, and paste code from your favorite technical books directly into your application.

The perks don't stop there, you can get exclusive access to discounts, newsletters, and great free content in your inbox daily

Follow these simple steps to get the benefits:

1. Scan the QR code or visit the link below

https://packt.link/free-ebook/9781804617403

2. Submit your proof of purchase
3. That's it! We'll send your free PDF and other benefits to your email directly

www.ingramcontent.com/pod-product-compliance
Lightning Source LLC
Chambersburg PA
CBHW080522060326

40690CB00022B/4998